KU-350-796

Robert Young, Ph.D., is a world-renowned microbiologist and nutritionist. He is head of the InnerLight Biological Research Center.

Shelley Redford Young is a licensed massage therapist and a chef specializing in optimum nutrition.

THE pH MIRACLE

Balance Your Diet,
Reclaim Your Health

Robert Young, Ph.D.

and

Shelley Redford Young

piatkus

PIATKUS

First published in the USA in 2002 by Warner Books Inc.,
1271 Avenue of the Americas, New York, NY 10020
First published as a paperback original in Great Britain by
Time Warner Paperbacks in 2003
Reprinted by Time Warner Books in 2005
Reprinted by Sphere in 2007
This edition published in 2009 by Piatkus
Reprinted 2010 (three times), 2011 (twice), 2012 (twice)

Copyright © 2002 by Robert Young, Ph.D.

The moral right of the author has been asserted.

The programme herein is not intended to replace the services
of trained health professionals, or be a substitute for medical advice.
You are advised to consult with your health professional with regard to matters
relating to your health, and in particular regarding matters
that may require diagnosis or medical attention.

All rights reserved.
No part of this publication may be reproduced, stored in a
retrieval system, or transmitted, in any form or by any means, without
the prior permission in writing of the publisher, nor be otherwise circulated
in any form of binding or cover other than that in which it is published
and without a similar condition including this condition being
imposed on the subsequent purchaser.

A CIP catalogue record for this book
is available from the British Library.

ISBN 978-0-7499-3981-6

Typeset in Caslon by M Rules
Printed and bound in Great Britain by
Clays Ltd, St Ives plc

Papers used by Piatkus are from well-managed forests
and other responsible sources.

MIX
Paper from
responsible sources
FSC
www.fsc.org   FSC® C104740

Piatkus
An imprint of
Little, Brown Book Group
100 Victoria Embankment
London EC4Y 0DY

An Hachette UK Company
www.hachette.co.uk

www.piatkus.co.uk

To Antoine Bechamp. If his profound voice and science had not been silenced, much of humankind may have been spared the worst aspects of the infectious and degenerative diseases of the twentieth century.

To our four wonderful and beautiful children: Adam, Ashley, Andrew, and Alex. And to our new son-in-law, Mathew, and brand new healthy pH Miracle grandson, CharLee.

And finally to our future—the children who are at the forefront of an ever-changing and challenging world. It is our hope that the message of the pH Miracle will be received within their minds and hearts and become the foundation for a healthier and happier world.

Acknowledgments

Dr. Robert O. Young

If I have seen further it is by standing on the shoulders of Giants.
—Sir Isaac Newton

This quote reflects how I personally feel about the men and women who have had a powerful impact on my life's work and mission. Their gifts of wisdom, knowledge, inspiration, and encouragement have lead me to The pH Miracle.

The life and research of French scientist and doctor Antoine Bechamp (1816–1908) set the foundation of my understanding on how matter can take on different forms and functions, a doctrine he referred to as pleomorphism. If it had been given the chance, his biological work might have revolutionized medicine with its profound insight into the nature of life, thus providing the cures for many sicknesses and diseases that science is still diligently seeking. His life's work opened my eyes and my heart and set me on my path of scientific research and discovery of the nature of sickness and disease and health and wellness. For this I am eternally grateful.

Modern Western medicine teaches and practises the doctrine of French chemist Louis Pasteur (1822–1895). The concept of specific, unchanging types of bacteria causing specific diseases made perfect sense at the time in my early training. Even though I do not now embrace the doctrine of monophorism, I am grateful to Pasteur for providing the basis of contemporary microbiological thought, which inspired me in my research on the nature of the germ and my quest for the truth of the matter.

I am richly blessed with a wonderful heritage of great men and women who make up in many ways who I am. My great-great-grandfather Brigham Young helped me understand the nature of matter, when he said that "matter cannot be created nor can it be destroyed." This very profound statement has echoed in my mind, as I realize that matter is eternal and can only take on different forms and functions.

My eternal companion, friend, and love of my life, Shelley, has been my river of inspiration and light. Through her love of people, especially children, and her creativity and intelligence, she has embraced The pH Miracle and made it possible for everyone to learn, understand, and live this New Biology.

Several years ago Dr. Neil Solomon and his wife, Frema, came to our home in Alpine, Utah, to evaluate the efficacy of my life's work. I was nervous and excited at the same time, because I knew he was a great scientist and researcher. Since their visit they have become good friends, as well as supporters of my work. In a world where scientific change is not necessarily received with open arms, I am grateful to a medical researcher who has had the courage to express openly the importance of this work and the need for continued scientific investigation.

I will never forget the day I received a call from bestselling author James Redfield, telling me he had given a copy of one of my previous books to Diana Baroni, an editor at Warner Books. He told her that she needed to read the book and consider publishing the author. Without the help of James and his belief in and personal experience with The pH Miracle, you would not be reading this now.

We then had the wonderful blessing of meeting Diana Baroni. We traveled to New York City, believing we were going into a meeting with a real corporate type. What we found was a very personal, caring, knowledgeable, and delightful person to work with.

Diana then introduced us to Colleen Kapklein, who became our partner in tailoring The pH Miracle into a beautifully simplistic gift to the world. Our thanks and gratitude to Colleen for her many hours of dedicated work.

And finally and foremost to my creator, who is the giver of all good gifts, the breath of all that is living, and who gives meaning and purpose to my life and mission. And who has also taught me the true nature and meaning of the blood and the anatomical living beings that make up all organized matter. In his words, "Life and death are in the Blood and from dust you are and to dust you will return."

Shelley Redford Young

It has been an incredible journey for me to ride alongside such a gifted and giving messenger. There is no way I can adequately express my gratitude to Robert, for his support, and his dedication and perseverance in courageously bringing this truth and light to a suffering world. He is a true servant. His contribution to mankind has made it possible for me to live out my most heartfelt desires—for people to be well, whole, and happy. It has been a rich blessing to share this healing wisdom all over the world, with Robert.

I want to express my deepest appreciation for our sons, Adam, Andrew, and Alex; and our beautiful daughter, Ashley Rose. They have been understanding, supportive, and patient in our absence, and have become devoted students and examples of the New Biology. Thank you for the immense joy you have brought into our lives!

Among dear friends whose support has been especially meaningful to this work, I want to include extended family members on both the Young and Redford sides who have encouraged me to produce recipes and get them "out there" into the hands of the public. You know who you are, and I love you!

I am most grateful for the recipe contributions from some gracious healers: Maren Hale, Angelique and Chrystyanna Queensley, and Lonnie Birdsall. Thank you for creating recipes that heal from the inside out.

Finally, to good true friends, Russ and Mary Anne Green (The Super Greens). Thank you for your never-ending love, encouragement, and support through this great roller coaster called life.

In love and InnerLight, and the Healing love and light of our Creator, who feeds us all,

Dr. Robert O. Young and Shelley Redford Young

Contents

Foreword

Another diet book?

No, far from it. This is a medically orientated book that argues, and justifies, the need for considered and sensible eating as an effective means of health control. Robert Young and Shelley Redford Young have simplified what is a complex matter and given the reader, whether professional or lay, a step-by-step guide to nutrition and diet.

I was most impressed by the practicality of the book. Simple urine or saliva tests that are performed at home can help govern dietetic changes. The authors have followed this up by providing simple and nutritious recipes and diet plans which are so often missing from health books. The guidelines for detoxing, re-alkalining the body and creating a simple-to-follow nutritional routine makes this publication unique amongst its peers.

Although some of the concepts portrayed within the book fly in the face of standard, pharmacologically led training, Young's research supports his principles and therefore challenges even the most sceptical of readers. The authors focus very much on the need for balance, particularly between acid and alkaline. This principle blends very well with the Eastern philosophies of yin and yang or vata/pitta/kapha and the book creates a Holistic feel throughout. It is not easy to scientifically justify a lot of the ancient and traditional lessons regarding health but the study of acid-alkaline balance has brought a scientific tangibility to traditional health techniques.

Doctors received less than two hours of nutritional training when I went through my six years of medical school. Since qualifying I have been fascinated by the array of advice that can be gleaned from nutritional and diet books. *The pH Miracle* would, or rather should, in my opinion, be a pre-requisite for anybody in the health field. Orthodox medicine, Western ideas of diet and nutrition and the immense power that the food industry has through advertising over the way we eat has to be altered sooner than later. It is arguable, and the authors do so very well, that many diseases stem from deficiency or food-related toxicity. The human body is capable of leading a healthy existence for over a hundred years and yet we struggle with ill health well before our allotted three score years and ten. Medicine is geared towards the profitability of allowing illness to occur and then providing drugs to reduce symptoms. The time has come to re-evaluate, and to establish correct nutrition, acid-alkaline balance and other lifestyle changes as a fundamental right owed to our consciousness (or soul) which inhabits our physical body.

This readable, organized and well-written discussion of an established but poorly promoted metabolic diet is an invaluable resource for those of us determined to maintain good health or repair from illness. I strongly feel this book should be integrated into medical colleges and feel privileged to have been asked to offer a validation to one of the finest books in its category. I, for one, will dine healthy and well because of it.

Dr R. Sharma © 2001

Dr Sharma is medical director of The Diagnostic Clinic in London.

Part I

THE
THEORY

Chapter 1

The New Biology of Health

A new day is dawning. A day of truly holistic health, vitality, and well-being. A time of energizing our very cells, maximizing the life force of our bodies. An era of naturally lean and strong physiques! This book is your key to letting that light into your life. It first helps you understand the revolutionary science I (Rob) call The New Biology, then explains the practical ways to change your diet and unleash the full power of your body to heal itself.

In truth, the light came over the horizon more than a hundred years ago. But the radical work of some great pioneering scientists has been overlooked by a mainstream medical model so deeply involved with its own myths that it was blind to larger truths. Until now. Now we're going to peel off the blinders and look at an entirely new model of human health—and disease. Working from decades of exhaustive research—my own as well as the work of those early (ignored) pioneers—we'll discover the basic pathways to illness and wellness.

You'll get the science, of course, but also the practical applications that will allow you to put the science to work for you.

It's all about balance. The universe operates by keeping opposites in balance, and the universe contained within your body is no exception. When imbalance occurs, we get the signs of disease: low energy, fatigue, poor digestion, excess weight, foggy thinking, aches and pains, as well as major disorders. This book is about reclaiming balance (health): energy, mental clarity, smooth operation of all body systems, clear, bright eyes and skin, and a lean, trim body. With the program in this book, all that will be yours within weeks.

All you have to do is take a look around you to see that most people with our modern lifestyles are suffering from imbalance. They are obese. Tired. Prematurely aging. Perhaps you are too. Chances are at least one person you love is suffering from one of the top three killers in the United States—heart disease, cancer, or diabetes. When I ask the audience at one of my lectures how many of them have a family member with one of these big bad three, 70 to 80 percent raise their hands. In fact, half of us will die from heart disease or diabetes. A third will die from cancer.

If you're like most people in our culture, at this point you're thinking, Well, you have to die of *something*. We've forgotten that it is natural to die—and to *live*—*healthy*! In fact, *it is your birthright* to *live* healthy, right up until the day you die. Making that vision a reality is the great gift of this program. And believe me, it has nothing to do with mapping the human genome, cutting-edge medical technologies, or even more powerful—and dangerous—pharmaceuticals. The good news is, the answer is much, much simpler than all that. And available right here and now. Today.

The obvious clue is right there in the top three killers (cancer, heart disease, and diabetes). All are directly linked to diet. In fact, *eight* of the top ten causes of death in the USA today are (not to mention that diet is obviously the cause of the obesity of 60 percent of Americans). Eating the proper foods and getting the best nutrients, *in balance*, will help you avoid all that—along with the misery and poor quality of life that so often precede death, sometimes by decades. The simple secrets to finding the right combinations are what The pH Miracle is all about.

Even mainstream medicine agrees: "Foods contain nutrients essential for normal metabolic function, and *when problems arise, they result from imbalances* in nutrient intake and from harmful interaction with other factors. For . . . adult Americans who do not smoke and do not drink excessively, *one personal choice seems to influence long-term health prospects more than any other—What We Eat*" (from the 1988 "Surgeon General's Report on Health and Nutrition": emphasis added).

Since that landmark report from C. Everett Koop (and even before, depending on the circles you run in), Americans have been following one diet craze after another not only in an attempt to drop the excess baggage most of us are carrying around but also as a way to find our way back to good health. The result? We (as a country) are fatter than ever. And we're certainly no healthier.

The problem is that while diet *is* the key not only to slimness but also to overall health and well-being, it has to be a diet that properly balances our body chemistry. But each and every one of the diets Americans have tried—including not just the "average American diet" but also such supposedly healthful regimens as low-fat diets, the food pyramid, and vegetarianism—create wildly *imbalanced* body chemistry. Even if one way of eating does get rid of extra pounds (usually temporarily) or lower cholesterol levels or lessen digestive trouble for some people, it does not fulfil the promise of simple good health. We're so used to our modern medical machinery, in fact, that good health seems anything but simple. I'm here to tell you: It *is* simple.

Forget cholesterol counts. Forget calories and fat grams. Forget blood pressure, blood sugar, hormone levels, or any of the other markers of health you're used to at the doctor's office. It turns out that the single measurement most important to your health is the pH of your blood and tissues—how acidic or alkaline it is. Different areas of the body have different ideal pH levels, but blood pH is the most telling of all. Just as your body temperature is rigidly regulated, the blood must be kept in a very narrow pH range—mildly basic or alkaline. The body will go to great lengths to preserve that, including wreaking havoc on other tissues or systems.

The pH level of our internal fluids affects every cell in our bodies. The entire metabolic process depends on an alkaline environment. Chronic overacidity corrodes body tissue, and if left unchecked will interrupt all cellular activities and functions, from the beating of your heart to the neural firing of your brain. In other words, overacidity interferes with life itself. It is (as we'll see in more detail in the next chapter) at the root of *all* sickness and disease.

If that's not enough to get you interested in balancing your body pH naturally, nondestructively, keep this in mind: Overacidity is also what's keeping you fat (more about that later).

The goal, then—and what this program allows you to do—is to create the proper alkaline balance within your body.

The way to do that is by eating the proper balance of alkaline and acid foods. That means 80 percent of your diet must be alkalizing foods, like green vegetables. (That percentage will go down somewhat once you've successfully rebalanced yourself.) That leaves much smaller portions of acid foods (such as meat and grains) on our plates. In addition, carefully chosen, high-quality supplements will help you achieve and maintain pH balance. You'll learn more about all this in detail in later chapters.

That's it. That's all there is to it. Get at least three-quarters of your plate covered with alkalizing food, use the supplements as explained in this book, and you'll be good to go.

The following chapters get into the details of *why*, and the remainder of the book shows you specifically *what* and *how*. You'll get the benefit of my own exhaustive research and the work of the great minds that have trodden this road before me, and you'll also get Shelley's real-life, practical expertise in how to make all this happen for you.

Bottom line: This program will bring you increased quality and quantity of life. I guarantee you'll see immediate improvement. Your energy will increase, you'll find new mental clarity and powers of concentration, you'll build strength and stamina, and you'll lose excess body fat while increasing muscle mass. You'll have bright eyes and clear skin. You'll look better. You'll improve your athletic performance. Your entire body will function more efficiently. Whatever health challenges you've been facing will improve and most likely evaporate altogether. In short, you'll regain all the effortless energy and wellness you thought was lost with your childhood. With the healing, wholeness, and rejuvenation that this program brings, you'll experience, possibly for the first time in your life, vibrant personal health.

Over and over again, we've witnessed joy, relief, and renewed peace of mind in those who have turned a serious or chronic illness around, those who lost weight they had been battling for years, those whose cholesterol levels dropped, skin cleared up, itching stopped, energy returned. I've seen people who no longer required insulin injections for their diabetes. People whose aches and pains disappeared. Even people who had been diagnosed with cancer whose tumors vanished and who were pronounced cancer free. We hear frequently from people who haven't been able to work for months or years who are returning to work, people whose allergies are letting up, people whose infections clear. People who are well, whole, and energetic again. People like Sharon and her husband. I'll let her tell their story:

I was up to a record 192 pounds. I was depressed and tired and had attempted almost every diet known to humankind. I would lose weight and then gain it right back when I returned to my old eating habits. I became more and more discouraged. I was sick and tired of being sick and tired—and fat! I knew it was time to make some major lifestyle changes.

So, once again I stepped onto the diet roller coaster by going to a local weight loss organization. It worked for the first few months. I lost thirty-two pounds, traded in my diet sodas for water, and started to exercise. I thought I might really have the problem licked this time.

At this same time, my husband applied for a life insurance policy. The findings of the company's physical exam scared us to death. His cholesterol was practically off the charts at 340. He had fatty growths in his shoulder. He was at his heaviest ever, too: 227 pounds. For someone who is five-nine, that's quite a load to carry around. The report made him out to be a walking time bomb. An accident waiting to happen. Just like me, he realized he needed to make some serious changes. But he was confused about what and how.

Before he could take action, we moved into a new house and went on a vacation. I went back to my old eating habits, stopped exercising, and regained twenty of the thirty-two pounds I had lost. When things settled down again and we realized we had to take action, that's when The pH Miracle came into our lives.

We shared the same first thought: How on earth could we change our lifestyle so drastically? And the same second thought: We had to try. That night we committed to the program—and immediately headed over to a favorite local restaurant to pig out on our last barbecued pork ribs and mud pie. The next morning we changed our diets—and changed our lives. We've never looked back.

My husband's cholesterol dropped to 242 in six weeks, and is now under 200. He had been experiencing arthritis in the hips and had been thinking about surgery. But the pain has disappeared! For the first time in my life, I'm not depressed. I'm no longer a compulsive eater; no longer do I eat until I get sick—and then eat some more. I'm not a slave to my addiction to sugar and breads. Now the foods I crave are avocados and soybeans. We both would still like to lose a few more pounds, but we don't worry about it anymore. We know our bodies will reach an optimum weight. And we'll never regain it because we no longer crave the foods that made us sick, tired, and fat in the first place.

It turns out the weight loss is almost a side benefit. What we've really gotten out of this is feeling better. Our minds and bodies are stronger. We no longer feel sick and tired all the time. Even our relationship has benefited. We have a more fulfilling life. We can more fully enjoy the best things in life—mainly, our kids.

Speaking of our kids, we are now gradually working them into the program, too. They've been raised on fast food and pizza, sweets and fizzy drinks. And they've been addicted to sugar, bread, and meat—and been tired, obese, and depressed as a result, just like so many of their generation. Who would have thought they would ever be drinking green vegetable juice for breakfast? We feed them healthy meals, and they take supplements as well. They still have some sugar now and then, but we know this is not an overnight process. The whole family is headed in the right direction.

Robert Louis Stevenson wrote: "There will come a time when we will sit down to the banquet of our consequences."

We were. And we are. I like this way much better.

Besides the huge volume of tales like this one—which I'll share more of throughout the book—I've seen the scientific proof of such transformations with my own eyes, though nothing is more persuasive than the radical changes in people's everyday lives. I've observed thousands of blood samples, including pH levels, from people all over the world, and seen the transformation that occurs—right down to the individual cells—when people change the way they eat.

HOW WE CAME TO EAT THIS WAY

All this healing occurs because people are willing to take responsibility for their own health and make the necessary changes. That was a mission Shelley and I were on for ourselves starting over twenty years ago. But it took quite a while to realize that what seemed in popular thought to be optimal was still leaving us tired, seemingly prone to every passing cold, and just not able to get the very best out of life. Eventually, through trial and error—experimenting on ourselves—we developed The pH Miracle program, and found, finally, the way to get fully, permanently, holistically healthy—physically prepared to live life to its fullest.

We had always been active, and on the lookout for ways to improve our health. By 1980, we were on what we thought was an ideal strictly vegetarian regime—mostly complex carbohydrates, along with some simple carbs (sugars), proteins, and fats. We were conforming to what seemed to be the best possible nutrition for our bodies and raised our children this way.

We ate the best of all foods sold in our favorite health food shops. Yet still we felt that afternoon fatigue. Our athletic endurance seemed to have plateaued, and we weren't getting any further improvements. We didn't have the strength and stamina we wanted. We were slowly but surely getting one common sign of aging after another. And it all seemed to be getting worse as we grew older.

What we didn't know then was what havoc all this was wreaking on our body chemistry, how it dangerously acidified our bodies, inviting all the disastrous consequences explained in upcoming chapters. Thank goodness the whole family had stayed generally healthy, but with what we know now we can see we were lucky. We were setting ourselves up for a lot of potential trouble.

It took a long time to realize what was going on—and what we could do about it. When we did figure it all out, what a dramatic difference! For one thing, though neither of us was overweight to begin with, we both shed pounds. I lost about twenty pounds—and four inches around my waist. Shelley dropped about fifteen pounds, and is now the same size she was in high school (nine). More important, we felt much better. Not only that, but I also could see astounding changes in our blood, including the pH levels. One good thing about the roughly fourteen-year journey we went through in developing this program was that it gave me time to come to some new (to me) laboratory techniques with which I could chart our progress (which I'll describe more fully later in the book).

Since we were essentially healthy to start out with, I was eager to test the effects of the dietary changes we were making on people facing serious health challenges, such as diabetes, cancer, obesity, heart disease, lupus, gout, and arthritis. Sure enough, in my studies I could see the same sweeping improvements in their blood. They all reported feeling better and losing weight. And in many cases, even the dire conditions healed or disappeared! Witnessing these amazing results, I knew we'd stumbled upon more than a way of eating that just happened to be right for us and our family. We'd found something we wanted to share.

We've been fully on The pH Miracle diet for more than five years now, and there's no doubt in our minds: We'll never go back. We've gained too much we'd never want to give up. Even as we get older, we keep feeling better. And we know we're protecting ourselves from the ravages of the diseases that claim so many of our fellow Americans. We've found a way to ensure that we'll be there for each other, for our work, and most of all, for our children, over the long haul. And able to truly enjoy every minute of it.

Good health should be second nature to us, but for many people it seems very difficult to maintain. The pH Miracle changes that for ever. It will empower you to regain what is rightfully yours, encourage you to take responsibility for your own well-being, and restore the great gift of wellness to you.

I know, there are lots of formulas out there for getting well. And many of them have good track records and positive reputations. Often, that's because they deal partially (perhaps unknowingly) with what you're going to learn here. Also, you can get a lot better without getting completely well, so you might think the job is done without ever reaping the greatest rewards. Once you experience the full, vibrant wellness that comes with this basic change to your diet, you'll know just how tremendous the difference can be. So we invite you to begin this journey and experience the results for yourself. Awaken to the light of a new day!

Chapter 2

The ABCs:
Acid, Blood, *Candida*

Before we get to the specific steps for transforming the way you eat—and the way you feel—I (Rob) want to give you a little background in the science supporting those steps. Think of it this way: I have good news and bad news—and I'm going to give you the bad news first. We'll get to the good news soon enough—the ways you can protect yourself from all this bad news, and the delicious food you can eat in the process—but for now I want you to understand some of the scientific foundation we've built The pH Miracle program on. Once you do, I think you'll see clearly why following the program is so important and can make such a profound difference in your life.

From the first chapter you know that the core concept most important to this program is the acid/base (alkaline) balance of your body, and particularly the blood, so we'll start by digging a bit deeper into that. Then we'll get into one of the nastiest consequences of an overly acidic body: the beasties that thrive there, including bacteria, yeast (the *Candida* of this chapter's title), fungus, and mold.

A IS FOR ACID

We're starting from the premise that of all the balances the human body strives to maintain, the most crucial is the one between acid and base (or alkaline). The body will go to great lengths to maintain the appropriate, slightly basic, nature of its blood. But it is all too easy and far too common for body tissues to become acidic. Such an imbalance sets the stage for chaos, opening the door to sickness and disease. Overacidification of body fluids and tissues underlies *all* disease, and general "dis-ease" as well. For one thing, it is only when it is acidic that the body is vulnerable to germs—in healthy base balance, germs can't get a foothold. Furthermore, acids are the expression of all sickness and disease. In short, good health requires a body in proper acid/base balance. Proper diet (like the one laid out in this book) is the only way to ensure that.

The relationship between acid and base is scientifically quantified on a scale of 1 to 14 known as "pH" (pronounced like the two letters). On that scale, 7 is neutral. Below 7 is acid and above it basic, or alkaline. Technically, pH reflects the concentration of hydrogen ions (positively charged molecules) in any given solution. But you don't need to understand the details of the chemistry here. Just know that these two kinds of chemicals—acids and bases—are opposites and when they meet in certain ratios, they cancel each other out, creating a neutral pH. In the blood, however, it takes about twenty times as much base to neutralize any given amount of acid, so it is better and easier to maintain than to regain balance.

B IS FOR BLOOD

Just as our body temperature must be maintained at 98.6°F or 37°C, our blood is ideally maintained at 7.365 pH—very mildly basic. (A mainstream doctor would accept up to 7.4, but that's problematic, as we'll see later.) You can also measure the pH of the urine and saliva, but the blood is the most important and needs to stay within the tightest range. Different areas of the body have different pH requirements anyway. For example, the blood and tissues should be slightly basic, but the lower bowel should be slightly acidic, and the urine slightly acidic or neutral. Saliva tends to be erratic. The pH of urine can provide the best estimate of what's happening in the body's tissues but it is not always accurate. Blood pH is more reliable, and thus a better indicator of internal conditions.

Physiological disease is almost always the result of too much acid stressing the body's pH balance, to the point where it provokes the body into producing symptoms of disease. (Disease can also be simply the toxic effects of an external source, but that is much more rare.) Symptoms can be the expression of that stress, but they can also be a sign of the body's effort to balance it. Depending on the level and extent of the stress, symptoms may or may not be obviously noticeable. The kicker is that excess acid is something we do to ourselves, thanks to the choices we make. The good news, then, is that once we recognize that fact, we can make different choices.

All the body's regulatory mechanisms (including breathing, circulation, digestion, and hormone production) work to balance the delicate internal acid/base balance. Our bodies cannot tolerate extended acid imbalances. In the early stages of the imbalance, the symptoms may not be very intense and include such things as skin eruptions, headaches, allergies, colds and flu, and sinus problems. As things get further out of whack, more serious situations arise. Weakened organs and systems start to give way, resulting in dysfunctional thyroid glands, adrenals, liver, and so on. If tissue pH deviates too far to the acid side, oxygen levels decrease and cellular metabolism will stop. In other words, cells die. You die.

So a declining pH just can't be allowed. To prevent it, when faced with a lot of incoming acid, the blood begins to pull alkaline minerals out of our tissues to compensate. There is a family of minerals particularly suited to neutralizing, or detoxifying, strong acids, including sodium, potassium, calcium, and magnesium. When these minerals react with acids, they create much less detrimental substances, which are then eliminated by the body.

Now, a healthy body maintains a reserve supply of these alkaline minerals to meet emergency demands. But if there are insufficient amounts in the diet or in the reserves, they are recruited elsewhere, and may be leached from the bone (as with calcium) or muscle (magnesium)—where they are, of course, *needed.* This can easily lead to deficiencies—and the many and varied symptoms that come with them.

That's just the tip of the iceberg. If the acid overload gets too great for the blood to balance, excess acid is dumped into the tissues for storage. Then the lymphatic (immune) system must neutralize what it can—and try to get rid of everything else. Unfortunately, "getting rid of" acid from the tissues turns out to mean dumping it right back into the blood, creating a vicious cycle of drawing out still more basic minerals from their ordinary functions and stressing the liver and kidneys besides. Furthermore, if the lymphatic system is overloaded, or its vessels not functioning properly (a condition often caused by lack of exercise), acid builds up in the tissues.

This imbalance in the blood pH leads to irritation and inflammation and sets the stage for sickness. Acute or recurrent illnesses result from either the body trying to mobilize mineral reserves to prevent cellular breakdown or emergency attempts to detoxify the body. For example, the body may throw off acids through the skin, producing symptoms such as eczema, acne, boils, headaches, muscle cramps, soreness, swelling, irritation, inflammation, and general aches and pains. Chronic symptoms show up when all possibilities of neutralizing or eliminating acids have been exhausted.

When acid wastes build up in the body and enter the bloodstream, the circulatory system will try to get rid of them in liquid form, through the lungs or the kidneys. If there is too much waste to handle, they are deposited in various organ systems, including the heart, pancreas, liver, and colon, or stored in fatty tissue, including the breasts, hips, thighs, belly—and brain. This process of acid waste breakdown and disposal could also be called "the aging process."

C IS FOR *CANDIDA*

The acid waste also sets the stage for the potentially devastating effects of a host of microscopic organisms in your body, starting with *Candida. Candida* is the Latin name for what is commonly known as a yeast in the human body but is really a kind of fungus. Yeast and fungus (and the closely related mold) are single-celled forms of plant life that inhabit land, air, and water. They are absolutely everywhere. For example, *Candida* is normally found in the gastrointestinal tract. We'd actually die without it. However, it can easily become drastically overgrown, causing a wide variety of symptoms, from annoying to chronic to fatal. This is the bug all too many women are familiar with through "yeast infections," and parents may have experience with if their infants ever had thrush (which is just *Candida* growing in the throat).

While mainstream medicine recognizes these and a handful of other medical problems stemming from yeast and fungus, the truth is that on the typical Western diet the vast majority of people develop out-of-control growths within their bodies—and the effects are disastrous. Actually, excess *Candida* is just one of the villains. We are living in a plague of "microforms," including yeasts, fungus, and molds as well as bacteria and viruses. Worse still, we are victimized not only by the microforms themselves, but also by their poisonous excretions, or "mycotoxins" and "exotoxins" (from "myco," meaning fungus, "exo," meaning bacterial, and "toxin," meaning, of course, poison). The microforms produce these acidic wastes when they digest (ferment, really) glucose, proteins, and fats—the same substances our bodies are looking to use for energy.

Candida and other microforms take advantage of the body's weaker areas, poisoning and overworking them. In an acidic environment, they basically get free rein to break down tissues and bodily processes. They live on our body's glucose, which they use for energy, and use our fats and proteins (even our genetic matter, nucleic acids!) for development and growth. These organisms are literally eating us alive! They then send their waste products (acids) out into the bloodstream, as well as inside the cells, further polluting the system.

Just to give you a little perspective on how daunting the potential damage is: Over the hundreds of millions of years that yeast, fungus, and mold have been on earth, they have developed into over half a million different identifiable forms. And they've undergone little genetic change. Apparently, they haven't needed to, because they are great opportunists and survivalists, perfectly suited to what they do. They can go from explosive growth to thousands of years of dormancy. (Living spores have been found in ancient Egyptian tombs only recently excavated.) Furthermore, there are more than a thousand toxins produced by yeast, fungus, and mold.

Bacteria, yeast, fungus, and mold do not themselves produce symptoms in the body—their toxic wastes do. Nor do they initiate disease. They only show up because of a compromised environment. As Rudolph Virchow wrote, "Mosquitoes seek the stagnant water, but do not cause the pool to become stagnant."

These organisms and their wastes contribute directly or indirectly to a huge list of symptoms. Most diseases, especially chronic and degenerative ones, follow microform overgrowth. Between the extremes of athlete's foot and AIDS are the yeast and fungus overgrowths underlying diseases such as diabetes, cancer, atherosclerosis (clogged arteries), osteoporosis, chronic fatigue, and more—including infections that appear to be transmitted person to person.

The general signs of overgrowth include pain, infection, fatigue, and body malfunctions including adrenal/thyroid failure, indigestion, diarrhea, food cravings, intestinal pain, depression, hyperactivity, antisocial behavior, asthma, hemorrhoids, colds and flu, respiratory problems, endometriosis, dry skin and itching, thrush, receding gums, finger/toenail fungus, dizziness, joint pain, bad breath, ulcers, colitis, heartburn, dry mouth, PMS and menstrual problems, irritability, puffy eyes, lack of sex drive, skin rash and hives, lupus, mood swings, hormonal imbalance, vaginal yeast infection, cysts and tumors, rheumatoid arthritis, numbness, hay fever, acne, gas/bloating, bowel stasis, low blood sugar, hiatus hernia, headaches, lethargy/laziness, insomnia, suicidal tendencies, coldness/shakiness, infections, over- and underweight conditions, chemical sensitivity, poor memory, muscle aches, allergies (airborne/ food), burning eyes, multiple sclerosis, malabsorption, and bladder infections (whew!). And that's not even including that general, just overall *bad* feeling so common these days. You can blame that on out-of-control microforms and their toxic acid wastes as well.

THAT WILL BRING US BACK TO . . . ACID

Microforms thrive in . . . acidity! They love to swim in their own waste products. They also love the low oxygen levels that come with acidity. On top of that, the wastes they produce are strong acids themselves. So just in case you needed further convincing about the importance of getting your body back to basics, try the mental image of your body swarming with mold and fungus.

Still, the good news is that eating properly and using

supplements wisely is all you need to do to let your body use and control the microforms it needs without risking the development of overgrowths or dangerous negative forms of them. Maintaining your acid/base balance through diet provides the optimal environment for *only* healthful levels of microforms.

Everyone needs to be concerned about microforms, even if they are not (*yet*) experiencing outward signs of overgrowth. That's because overgrowth happens in two stages. In the first, initial development phase, microforms grow in small colonies and, although they are most likely visible in the blood, are probably not detectable by physical sensations or symptoms. In the second, acute or chronic symptomatic phase, the complications and discomforts become obvious. It is now bad enough that your body is complaining, throwing out warning signs and pleading for help. A second-stage overgrowth may happen relatively quickly or take years to develop.

Even in this second, more serious stage, all you need to do to reverse it is to create an internal environment in your body that won't support the microforms. All that requires is balancing your blood and tissue pH with nutritional supplements and an alkaline diet like the one detailed in later chapters. Of course, it would be better still to stop them in their tracks before they ever get that bad, which is why anyone will benefit from this program.

When your body goes from acid back to base, yeast, fungus, and mold stop growing and revert to being benign. Their leftover toxins can then be bound up by certain fats and minerals and eliminated from the body.

WHEN THE FISH ARE SICK, CHANGE THE WATER

Think of your body as a fish tank. Imagine your cells and organ systems as the fish, bathed in fluids (including blood) that transport food and remove wastes. Then suppose I back up a car and put the tailpipe up against the air intake filter that supplies oxygen to the tank. The water becomes filled with carbon monoxide, making it acidic. Then I throw in too much food, or the wrong kind of food, and the fish are unable to consume or digest it all, so it starts to decompose. Toxic acid wastes and chemicals build up as the food breaks down, making the water still more acidic.

How long before the fish are goners?

You'd never do such things to the most ordinary of goldfish, yet we humans do the equivalent every day to our own bodies, our own blood, fouling them with pollution, excessive intake of food, acidic foods, and more. The fish are floating belly up, but it is as if we can't see them, or don't know what it means.

Now, back to our polluted fish tank. If you reached such a sorry state of affairs, what would you do? Would you treat the fish for the illnesses they would no doubt develop? No—you'd change the water.

Do your body the same favor. Change the water. Clean up the environment. Then keep it clean. The program in this book shows you how.

HISTORY, LOST AND FOUND

Classical biology, based on the work of Louis Pasteur in the late 1800s, relies on the idea that disease comes from germs that invade the body from the outside. But in studying the dazzling but shamefully overlooked work of Pasteur's contemporary Antoine Béchamp and his followers, including Günther Enderlein, Claude Bernard, Virginia Livingston-Wheeler, and Gaston Naessens, I've learned that, in an acid environment, bacteria and other microforms can *come from our own cells.*

Pasteur's "germs of the air" may contribute to illness, but they are not, contrary to popular belief, necessary for illness to occur. Their negative effects are simply added to the compromised environment already existing in the body.

Besides generating various microforms within our own bodies, we do also have them coming in through our respiratory system and intestinal tract (often via our food—but more about that later). Bacterial invaders *appear* to then grow in the body, wreaking their characteristic havoc. But what really happens is that their presence initiates a similar development in the bacteria already in the host—depending, again, on the environment.

An acid environment gives a big green light to this process. To contract an infection, you have to be predisposed to it internally. You have to have some of the bug already in your system, and you have to have the acidity to allow it to take hold. This is why some exposed people get a cold or any other bug, and some don't. Think for a minute of the flu epidemic of 1918. It ravaged the planet, killing about 30 million people worldwide. But it would occur in one home but not the one next door, one family member but not the next. Why? If you throw seeds on concrete, they won't grow. They have to meet fertile soil. So it is with germs. Even if they do get into your body, unless it is nice and acidic, they can't grow and multiply and make you sick—or kill you.

MANY FORMS

The other key fact about microforms is that they can rapidly change their form and function. Bacteria can change into yeast, yeast into fungus, and fungus into mold. With this work being overshadowed by Pasteur's, we have for more than a hundred years lost the critical knowledge that disease is a condition of our own inner environment, not something caused by attack from foreign entities.

This lost chapter of history reveals that there is something living independently in cells and body fluids that is capable of evolution into more complex forms. These elements are known as microzymas ("micro," meaning small, and "zyma," meaning being), and all living things contain them. Degeneration and regeneration both originate with the microzymas. All cells evolve from them to begin with. In the right circumstances and environment, microzymas evolve into more complex life forms, including bacteria and fungi. It is a two-way street: Bacteria can also de-evolve back to microzymas. Everything begins and ends with microzymas. What happens in between depends on the environment.

The ability of microforms to evolve, to change form *and* function, depending on their environment, is known as pleomorphism ("pleo," meaning many, and "morph," meaning form). My theory is that red blood cells do this too: They can de-evolve and then re-evolve into any kind of cell the body needs—bone cells, muscle cells, skin cells, brain cells, liver cells, heart cells, and so on. In a kind of parallel process, bacteria, yeast, fungus, and mold are morbid evolutions of healthy cells (including red blood cells, brain cells, and liver cells).

You are already familiar with one chemical example of pleomorphism: the transition of plain water to steam—or snowflake. The chemical structure doesn't change—it is still just H_2O—but the form does, depending on the environment.

Now, I bet you can guess what kind of environment spurs morbid changes in microforms in the human body. That's right: acidity. Microzymas don't always become bacteria, and bacteria don't always evolve into fungus, nor does fungus always become mold—it takes an acid environment. Harmful pleomorphic organisms do not, and cannot, evolve in healthy (alkaline) surroundings.

With a high-powered light microscope, a video recorder, and a printer, I have been able to record the evolution of pleomorphic organisms from rod-shaped bacteria (*bacilli*) to spherical (*cocci*), and ultimately into yeast and fungus and mold—and back again. Pleomorphism has also been seen in recent electron microscope pictures of animal tissue.

Dramatic experiments have demonstrated the extent of the transformations possible. For example, one type of amoeba (a single-celled organism) feeds on bacteria and another, a dysentery amoeba, eats rice. The two have unique DNA specific to their form. But an amazing thing happens if you flip-flop their diets, gradually switching the one amoeba over to rice, and the dysentery amoeba to bacteria: Their genetic material actually changes! They literally switch into each other. That makes microform pleomorphism even more profound than the change of a caterpillar into a butterfly, and all the more fantastic because it can happen quite rapidly, sometimes in a matter of seconds.

Pasteur's friends in high places, his showmanship, and his ability to basically market himself and his work started the ball rolling in favor of "the germ theory" all those decades ago, and mainstream medicine adheres to it to this day. That tradition is so strong, and the alternatives so revolutionary, that even something that is plain to see, observable with your own eyes, nonetheless goes unseen. My fervent hope is that this is, all too slowly, starting to change.

There is one more reason this lost history is so slow to be recovered. Just as microforms can evolve, they can also revert to their original state. For example, in beer, only a trace amount of the yeast initially added to a batch is present after the fermentation of the grain, and it is no longer visible to the unaided eye. Only the alcohol—simply a mycotoxin—is left. Where did the organism go? It isn't really gone, of course, it has just returned to microzymas. Similarly, every cancer tumor is surrounded by a pool of lactic acid—another mycotoxin—but the microform may or may not be there. So even those willing to look won't *always* find.

Those willing to look again, and with clear eyes, will be rewarded with the secrets to permanent health. We can heal ourselves by changing the environment inside our bodies. Potentially harmful invaders, then, will have nowhere to grow and will become harmless.

ACID IMBALANCE IS PERFECTLY NATURAL . . . WHEN WE'RE *DEAD!*

The chaos of acid imbalance and microform overgrowth is an entirely natural and orderly process when life is ending. The body automatically becomes acidic upon death. Once a body stops breathing, oxygen levels of course decrease, creating the anaerobic ("without oxygen") environment microforms thrive in (in addition to the acid they love).

Then these little buggers get down to work. Their one big job—one reason they are a part of the normal human body—is that they are the principal "undertakers" when we die. Those mycotoxins are designed to decompose our dead bodies. The microforms and their toxins are here to reduce us into our simplest component parts—back to microzymas. Biologists call it the carbon cycle. It's the literal meaning behind "ashes to ashes, and dust to dust." In less technical or poetic language: They are what make our corpses rot.

With microform overgrowth in overly acidic *living* bodies, that process is set in motion prematurely. Yeast and fungus start their takeover while we are still living. We are basically rotting inside. Fermenting. Molding. Take your pick!

Keep in mind, however, that there is nothing inherently bad about the microforms themselves. If anything, they are actually good. Cells all over the body must constantly break down and renew themselves to stay healthy and vigorous. Microforms are there to handle the recycling, so the garbage doesn't pile up.

GERM, SYMPTOMS OR DISEASE?

Unfortunately, Pasteur was confusing disease with its symptoms, and that central misconception has come down through the generations as scientific law. In reality, disease is a general, underlying condition, not the symptoms we diagnose. If germs are involved, they are themselves just symptoms of that underlying condition. Remember that germs come from within our cells, and that germs invading from outside the body can only contribute to a state of imbalance and stimulate secondary symptoms. What most people call disease is really just a collection of these secondary symptoms. Germs are really just the expression of the underlying disease condition (overacidity and microform overgrowth).

Over the last century or so, mainstream science has decided the precise causes of some "diseases." But many serious ones still seem to be something of a mystery . . . until you understand that no matter what the symptoms bothering you, the immediate causes are always the same: acidity and microform overgrowth.

COMMON PROBLEMS OF OVERACIDITY

Your body faces all kinds of breakdown if it is allowed to get too acidic or is forced to fight too hard for too long just to stay basic, or if it gets overgrown with noxious microforms. As I said before (and will no doubt say again), if you dig deep enough, these twin problems underlie just about anything that ails you. Here I want to look at some of the most common symptoms that result.

Weight

You can thank an overly acid internal environment for the excess pounds you are carrying around. In a defensive maneuver, the body creates fat cells to carry acids away from your vital organs to try to protect them. In one sense, your fat is saving your life! But that's why your body doesn't want to let it go. When you eat to make your body more basic, your body won't need to keep that fat around anymore.

Weight problems can also result from yeast and fungus interfering with the digestion of food. The nutritional deficiencies created can actually trigger your body to pack on extra pounds, in part because you are always hungry. More commonly, blood poisoned by mycotoxins goes to the liver to be detoxified—and that added stress distracts the liver from efficiently metabolizing fat and sugar.

The chaos in an imbalanced body will exhaust the adrenal glands, and the resulting low levels of energy contribute to weight gain. Another likely villain is fatigue of the thyroid gland—which controls the rate of metabolism. Cravings for sugar, outsized appetites, and low blood sugar levels all follow an overgrowth of harmful yeast, fungus, and molds in the body.

Taken together these patterns all make it easier to gain fat and harder to lose it. To top it off, poor digestion and possibly depression will develop or worsen, too. Those are only a couple of the wide variety of ways the chaos of imbalance can express itself, as it did for Tara (see box). Ironically, though she had tried to control her weight for years, it was when Tara finally changed the way she ate not to lose weight but to address the symptoms plaguing her that she finally did drop her excess pounds—without even trying.

TARA'S STORY

I've been on weight loss diets for as long as I can remember—at least since I was eleven years old. I would starve my body into submission, but as soon as I returned to eating anything close to "normal," I would gain back all the weight I had lost—plus a few extra pounds. Aside from my extra fat, I always considered myself healthy, strong, and energetic.

All that changed when, after a serious infection in my uterus, I had a hysterectomy for large fibroid tumors. I never felt as though I recovered from the surgery. I developed pain in my left breast. The doctors thought there was a lump, but after mammography and ultrasound decided it was just a cyst. Either way, it didn't explain (or relieve) my pain. I became fatigued to the point where, if I went to the grocery store one day, I had to spend the rest of that day, and all of the next day, in bed.

This went on for years. Doctors recommended antidepressants, but I refused them. What I was experiencing was certainly depressing, but I did not feel I was depressed. I have a wonderful husband, lots of interests, and many things I want to do with my life. We spent thousands of dollars pursuing many different avenues to restore my wellness, and eventually I was able to increase my energy from about 20 percent of normal to about 70 percent.

It was right around then that I ran across the Youngs' program. Within days we had all the supplements we needed and began to change the way we ate. We had been off caffeine, alcohol, meat, and dairy for quite a while, so sugar was the only big challenge now. The first few days were rough. I was feeling sorry for myself and missing dessert. What I really wanted, however, was my health! It was the first time I went on a "diet" not to change my weight but for health alone.

After four weeks, we happened to be at a friend's house where I noticed a scale. We no longer owned one, as it had become an instrument of torture for me during my continuous dieting to lose weight. I stepped on and could barely believe what I saw: I had lost twenty-five pounds! I knew my clothes were getting looser, but I didn't feel as if I was really dieting because I wasn't going hungry. It turns out my husband, too, had lost twenty-five pounds. He's had to buy new trousers with a four-inch-smaller waist.

Except for the extra weight, my husband thought he was pretty healthy—he went on the program to support me, and I will always be grateful for his loving help. But he got more out of it than he knew at the beginning. A blood test early on indicated a prostate imbalance, even though he had no other symptoms. Six weeks into the program, a repeat test showed all signs of the imbalance were gone!

We are both thrilled that our bodies are moving toward normal. Our energy levels are continually improving. Emotionally, we are more stable. Even under great stress, we don't overreact. I have no more pain in my breast, and the "granulomas"—little hard lumps on the skin—we both had are melting away. We are mentally more alert—and noticing how sleepy others seem. I think we spent most of our lives in a sugar fog.

We have been checking our urine with pH paper. When we started, it was 6.0 or less. Now we run around 6.8 to 7.0. Best of all, our friends are overwhelmed at the changes they see in us. They all say, "You look great!

What are you doing?" I tell them about our diet and supplements. But really I think about what we're doing as simply growing young together!

At the other end of the spectrum, the yeast and fungus produced within an overly acidic body can feed on your protein and other nutrients, interfering with the absorption of everything you eat by as much as 50 percent. This can cause you to become excessively thin, which is no healthier than being overweight. As you restore a healthy balance in your body's environment on a basic (alkaline) diet, you will begin to gain weight, then stabilize at your ideal.

Healthy bodies are not overweight or underweight. A healthy body naturally maintains its own ideal weight.

Allergies

The toxins produced by microforms within an overly acidic, oxygen-deprived body contribute significantly to what are commonly considered the symptoms of allergy. The mycotoxins severely stress the immune system, so it is constantly stimulated, overworked, and on edge. Imagine trying to clean your house while filth is constantly being tossed in through the windows! The result is chemical or nutritional sensitivities—allergic reactions (irritation and inflammation) to foods, airborne matter, or chemicals (including "hay fever," allergenic asthma [like Jennifer's—see box on page 36], environmental sensitivities, and food sensitivities or "allergies"). Soreness, swelling, watery eyes, runny nose, and eczema are all ways of eliminating acid toxins.

If you had no symptom-producing yeast and fungus, it would be impossible for you to have allergies—another reason to keep your body in acid/base balance.

Fatigue

Fatigue is probably the major symptom or complaint of an overly acidic body or a body overgrown with negative microforms (see box on page 36, "Jennifer's Story"). Microforms ferment sugar in our bodies that we would otherwise use for energy (then, to add insult to injury, they spew out acid waste as a result). It's as if you were trying to fill your car's petrol tank as someone else was siphoning off the petrol. Without the energy it needs to keep going, your car won't get very far or perform very well. The toxins produced in an acidic body reduce the absorption of protein, minerals, and other nutrients, which in turn weakens the body's ability to produce enzymes and hormones and the hundreds of other chemical components necessary for cell energy and organ activity. This also interferes with the reconstruction of cells and other necessary components of energy production. The result is fatigue, poor endurance, an inability to add muscle tone, and general weakness (as well as unwanted body weight changes and illness). The pancreas, liver, and adrenal glands, which play major roles in controlling energy levels, are all susceptible to the negative effects of mycotoxins.

Microforms—yeast, fungus, and mold—also rapidly deplete your supplies of the B-complex vitamins, iron, and other minerals. That alone could cause fatigue. Another result of microforms draining off your body's nutrients is that you have rapid drops in blood sugar (glucose) levels, which again create fatigue, poor endurance, and weakness. Yeast and fungus unbalance the process that controls the water and mineral content of cells (electrolyte balance), which is necessary for cell activity, thus impeding the normal flow of energy.

Fatigue vicious circle number one: Low energy levels encourage overgrowth of harmful microforms.

With all this, you won't be surprised to learn that microforms are the major players in chronic fatigue syndrome, which may involve yeast and fungus damage to nerve tissue and interference with nerve transmission thanks to a breakdown of neurotransmitters. In addition, acidic mycotoxins can strip away the myelin sheath that coats and protects nerves and enables transmission of impulses.

To give you just one example of the mechanisms behind the fatigue phenomenon, let's look at one mycotoxin, known as acetaldehyde, which is created from the fermentation of sugar (and by the end of this paragraph you'll have yet another good reason to stay away from sugar). Acetaldehyde can reduce strength and stamina, cause excessive fatigue, cloud thinking, and block ambition. One way that happens is that it reduces the absorption of protein and minerals, decreasing the ability to produce crucial enzymes and hormones. Another way is that it destroys essential enzymes, reducing cell energy. Third, acetaldehyde directly destroys neurotransmitters, which are chemicals responsible for completing all nerve impulses. A fourth is that it binds to the walls of red blood cells, like molecular glue, making them less flexible and therefore less able to get into and through the capillaries of the circulatory system. That leads to starvation and oxygen deprivation in the tissues. Furthermore, the liver converts acetaldehyde into alcohol. That process depletes the body of magnesium, sulphur, hydrogen, and potassium, thus reducing cell energy. And of course the alcohol itself has negative effects. It can actually produce the same symptoms as being drunk, making you disoriented, dizzy, or mentally confused. Vicious circle number two: The less oxygen there is in the body, the more alcohol is produced.

JENNIFER'S STORY

Even when I first began playing volleyball on my high-school team, asthma was a constant annoyance. The suffocating feeling of asthma always scared me, but I didn't let my breathing problems get in the way of the game I immediately loved so much.

But a year later, while I was playing for a nonschool team in the spring, my energy really began to flag. I thought my inhaler might be bothering me, so I went in for a checkup with my doctor. He gave me an extra inhaler to take in the morning along with the one I had been using before games.

So I wouldn't be dependent on inhalers, my family helped me change my diet to eliminate all sugar and dairy products. This helped a little with the asthma. But still, if I didn't use the inhaler before I exercised, I'd get an attack. I seemed to be tired all the time.

By the next spring, I was sleeping fourteen to fifteen hours a day. I could only handle one class at a time at school. I felt as though I was losing my bond with my friends, because on weekends I was too tired and depressed to get together with them. I had to use part of my summer to make up the classes I couldn't attend during the year.

It was another year before I found out about The pH Miracle, and how yeast in my system would cause extreme fatigue, sleepiness, and depression. I did a three-day fresh green vegetable juice fast, began taking supplements, and started eating whole green vegetables. Later I added turkey, grains, sprouts, and yellow vegetables.

I regained my strength and had more energy than before. I felt as though I could make something of life again. I lost five pounds the first week. I was able to complete a full volleyball session and join a volleyball club. I'm back to a full day of classes at school. I work out at the gym, work part-time, and have energy and vitality to be with many friends again. Best of all, my asthma is gone!

If it weren't for this program, I think the only way I could participate in volleyball would be to read about it on the sports page. Now when I read about my school's team in the newspaper, I often see my name as the high scorer!

Mood Disorders and Neurological Imbalance

As with Jennifer's experience, depression and other mood disorders are another result of acidic bodies and microform overgrowth, and they have reached epidemic proportions.

The usual pathway is mycotoxin interference with the production of coenzyme A. Coenzymes combine with other compounds to make enzymes, which are necessary in almost every bodily process, including those of the brain and nervous system. When coenzyme A decreases, conditions such as depression, anxiety, panic attacks, irritability, mood swings, and PMS often appear or worsen. Other symptoms come simply from being poisoned: paranoia, not being in total control of one's actions, knowing the right thing to do but being unable to do it, mental incompetence, and a variety of other behavioral, emotional, and psychological disturbances. Another variation is hypochondriac-type reactions or neurotic behavior and emotional instability. People may be very aware of their behavior, and be miserable about it, yet still be unable to control themselves because the toxins stay in their bodies.

Microform overgrowth causes an increase in acid, which leads to other neurological fallout, too, including headaches, migraines, inability to concentrate, memory problems, confusion, dizziness, that "fogged-in" feeling, and even MS-like symptoms such as slurred speech and lack of muscular coordination.

Sugar Metabolism

Low blood sugar (hypoglycemia) and high blood sugar (diabetes) are rampant today, and devastate a lot of lives. They both stem from—surprise!—microform overgrowth and the fermentation of our sugars thanks to these yeasts, fungi, and molds. Mycotoxins and exotoxins penetrate, overwork, and poison the pancreas, liver, and adrenal glands (among others) and disrupt sugar metabolism.

For example, pancreatic cells are directly poisoned and destroyed by the mycotoxin alloxan. The pancreas not only produces enzymes for digestion, but also insulin, the hormone that controls blood sugar and allows it to enter cells and produce energy. A deficiency in the hormones that control insulin is one pathway to hypoglycemia.

Yeast and fungus also feed on hormones, and can cause deficiencies that way. They also feed directly on blood sugar, disrupting the body's balance, lowering sugar levels, and overworking the liver as well as the pancreas and adrenals. Toxification of the liver interferes directly with glucose production.

Other Nagging Symptoms

I hope by now I've made it clear that acidification and overgrowth of negative microforms in the body are root causes of every symptom, illness, and disease (which are really only symptoms themselves). Here I just want to mention some of the common symptoms that are a direct result of overacidification and yeast and fungus overgrowth. These include vaginal infections, menstrual difficulties, impotence, infertility, prostatitis, rectal itch, urinary tract infections, and urgency and burning with urination. Respiratory manifestations abound: In addition to allergies (see page 33), congestion, excess mucus, postnasal drip, frequent clearing of the throat, habitual coughing, sore throats, earaches, and even asthma and bronchitis are often the result of fungus. So is the habit of catching everything that's "going around"—coming down with every cold and flu. The skin also has a variety of ways of manifesting microform overgrowth: Athlete's foot, jock itch, skin rash, hives, moles, birthmarks, dry browning patches, ringworm, rough skin on the sides of the arms, fungal nails (nails are modified skin), acne, and even skin tumors.

INFANTS AND CHILDREN ARE
PARTICULARLY SUSCEPTIBLE

Most childhood infections and symptoms are caused by negative microform overgrowth, including nappy rash, thrush,
ear infections, tonsillitis, colic, constipation, and diarrhea.
Even the poorly understood Sudden Infant Death Syndrome
(SIDS) is linked. As children become older, conditions like
attention deficit disorder (ADD), hyperactivity, aggressive-
ness, irrational behavior, poor self-esteem, learning
disabilities, and short attention spans can develop.

A mother's acidity or overgrowth of negative microforms
will certainly affect her newborn, and mother and child often
have similar problems.

HOW ACIDIC ARE YOU?

You can check your pH levels at home with paper pH strips,
available at many pharmacies or from tropical fish retailers in
the UK. You can also source them via the Internet. In addition you could also use diasticks that diabetics use to test
their urine.

The strips, which are relatively inexpensive and should be
easy to find, test the pH of your saliva or urine. The pH of
saliva is much more variable, so you are better off testing
your urine. Urine pH changes too, in response to what you
eat, so first thing in the morning, after you've fasted
overnight, is the ideal time to test.

The strips change color to indicate acid or base, and are
lighter or darker depending on the intensity of the reading.
They come with a color chart to help you translate the color

into a number. In most cases, you're looking to turn your strip a medium green—not dark, or bluish, meaning too basic, or light or yellowish, meaning too acidic.

So test yourself to see where you are right now, and then retest to keep tabs on your progress. You can also see for yourself the effect of meals on pH, by regularly testing with pH strips. Though the results are not definitive, you will at the very least be able to see trends. Test yourself after the basic meals like the ones described later in this book, and compare the result to results you got on your usual diet.

HOW OVERGROWN ARE YOU?

If your urine or saliva (or blood) are acidic, it's a safe bet that you have microform overgrowth. The simple fact is, most people do.

Live blood analysis more directly detects overgrowth. In a standard lab evaluation, drops of blood are basically dried onto a slide to be examined under a familiar bright-field microscope, where many of these negative microforms cannot be seen. Live blood analysis, by contrast, looks at unaltered live blood, under special dark-field, phase-contrast microscopes. The high-powered microscope can magnify objects up to 28,000 times, so you can clearly view bacteria, yeast, fungus, and mold in exact detail in the blood. You can also see red and white blood cells, crystallized microforms, mycotoxins, cholesterol, metals, blood clots, undigested fats, and many other things—all in *one* drop of live blood! Bottom line: Though it can also provide a lot more information, live blood analysis gives you a plain view of how crowded your blood is with undesirable microforms.

Finally, after years of researching German techniques in dry blood analysis, I've developed a test called the Mycotoxic/Oxidative Stress Test (M/OST), which involves a small amount of blood allowed to dry and clot on a microscope slide. Under the microscope, blood from healthy people forms a standard pattern—a dense mat of red areas interconnected by dark, irregular lines. The blood of people under mycotoxic/oxidative stress—meaning excess acidity and microform overload and the resulting harmful wastes—has a variety of characteristic patterns that deviate from that norm. One common (and visually striking) abnormality is the presence of "clear" or white areas interrupting the standard pattern. The extent and shape of the clear areas reflect the symptoms that are likely to arise as a result of excess acidity and overgrowth. That is, the pattern of the blood reveals not only the presence of microform overgrowth and excess acidity but also the particular ways that overgrowth is affecting the individual. Certain patterns match certain symptoms, such as diabetes, arthritis, atherosclerosis, and cancer.

In the end, however, getting all the details about your exact situation isn't necessary (though witnessing a live blood analysis may be motivating like nothing else!). Anyone on the standard Western diet is, to a greater or lesser degree, imbalanced—acidic. If you have any symptoms, you can be sure you are imbalanced. On the other hand, if you follow the program outlined in this book, doing what you know is right for your body, you can rely on your body to handle the complex details of self-repair. Your results, in how you look and feel, will speak for themselves. Freed from acid overload, you'll be symptom-free, full of energy, and mentally wide awake. You'll also reach your body's own best, healthy weight.

WHAT CAUSES IMBALANCE AND OVERGROWTH?

Overacidity and microform overgrowth are inextricably linked. Microforms are a major source of acid in the body. Acidification creates a comfy environment for microforms. We predispose ourselves to both conditions through various stresses. The main one is poor diet, although chronic toxicity from external sources and other physiological stresses (including poor digestion, more about which follows in Chapter 3) play roles as well. Emotional upheaval, negative thinking patterns, and other psychological stress also contribute.

This is what I call the cycle of imbalance. And a vicious cycle it is, going round and round and round once it gets under way—unless you step in and take action.

First comes something that disturbs your body in some way, be it poor diet, polluted environment, negative thoughts, spiritual distress, or destructive emotions. Whatever it may be, that initial physical or emotional disturbance starts acidifying your body and disturbs your very cells. Cells work to adapt to the declining pH of their compromised environment. They break down and evolve to bacteria, yeast, fungus, and molds. These in turn create their waste products—debilitating acids—which further pollute the environment. That in itself is a disturbance to the system, and in this way the whole cycle keeps rolling along.

THIS PROGRAM CHANGES ALL THAT

No matter how you got there, or how deep you're in it, a healthy, plant-based diet and low-stress lifestyle will keep you in acid/base balance and housing only helpful micro-forms. Eating the right kind of food is the single most important thing you can do for yourself and your health.

This program will restore health, harmony, and balance to your body through a diet based on alkalizing vegetables, sprouted and soaked nuts and seeds, essential oils, and low-sugar fruits. You'll experience a new level of wellness, energy, and mental clarity. Normalizing the blood and tissue pH will reduce the amount of symptom-causing microforms in the body—and thus reduce symptoms. With this program, you can also have the lean, trim body you've always wanted. As you get back to basic, the body naturally begins to seek its own ideal weight.

The thousands of blood samples I've studied from all over the world reveal the amazing cellular changes that occur with diet changes. As a person eats more alkalizing foods, espe-cially raw vegetables and greens, I see extreme improvement in red blood cell integrity, oxygenation of the blood, and levels of negative microforms. The same methods we covered for measuring acid imbalance and overgrowth will confirm for you that you are on the right path once you start eating according to this plan. Of course, you won't need the tests to tell you—the disappearance of your symptoms and your restored or renewed vitality will tell you all you really need to know.

Chapter 3

D Is for Digestion

Good digestion is critical for good health. The human body requires efficient digestion and proper elimination in order to maintain well-being and energy levels. Yet there is no more common physiological malfunction in humans than indigestion, in all its many and wonderful forms. Consider this: Antacids (for taming just one of those forms of indigestion) are the number-one over-the-counter remedy in the United States. Yet when we tolerate or ignore these conditions, or mask them with some pharmacy chemicals, we're missing the urgent messages our bodies are sending us.

We must listen. The discomfort *should* serve as an early warning system. Indigestion underlies most disease and its symptoms because digestive disturbance supports microform overgrowth and the resultant toxins. (It's another vicious cycle: Overgrowth of yeast, fungi, and mold also contributes to indigestion.) Poor digestion promotes an acidic bloodstream. Furthermore, we can't be properly nourished if we're not properly digesting our food; without proper nourishment, we can't be fully and permanently healthy. Finally, recurrent or chronic indigestion on its own can be deadly, gradually impeding intestinal function, which can go unnoticed until serious conditions like Crohn's disease, irritable bowel syndrome, and even colon cancer exist.

1, 2, 3

Digestion actually has three key parts, and all of them must be in good working order to maintain good health. But problems are common at each of the three steps along the way. First, there's indigestion, beginning in the mouth and continuing in the stomach and small intestine. Second, there's reduced absorption in the small intestine. Third, there is lower bowel constipation, which can show up as diarrhea, infrequent bowel movements, fecal impactions, bloating, or foul gas.

Here's a tour of your digestive tract to help you understand how these types link and overlap. Digestion actually begins even as you are chewing your food. In addition to the tearing and grinding of your teeth, saliva also begins to break down the food. Once the food reaches the stomach, stomach acid (a heavy-duty substance) continues the breakdown of the food into its component parts. From there the digested food moves into the small intestine for a long and winding journey (humans have up to 27 feet or 8 meters of small intestine), during which nutrients are absorbed for use in the body. The next and final stop is the colon, or large intestine, where water and some minerals are absorbed. Then whatever your body hasn't absorbed you excrete as waste.

It is a neat and efficient system when it is working right. It is even pretty resilient. But we habitually overtax our digestive system with low-quality food pretty much devoid of nutrients—not to mention the stress most of us live with—to the point where in the vast majority of Americans, digestion is simply not occurring as it should. And that's before you factor in the microform overgrowth and acidic environment!

"FRIENDLY" BACTERIA

That's the basic anatomy. The other crucial component of the human digestive system you need to understand is the bacteria and other microforms, in large quantities, in permanent residence. Until our lifestyles and habits interrupt, these *friendly* bacteria, known as probiotics, exist within us to actually help us stay healthy. They are indispensable—essential not only for health, but also for life itself.

Probiotics maintain the integrity of the intestinal wall and the internal environment. They prepare food for absorption and assimilation of nutrients. They help provide proper travel time for digested food passing through, to allow maximum absorption of nutrients but also swift elimination. Probiotics excrete a variety of beneficial substances, including the natural antiseptics lactic acid and acidophilin, which also aid in digestion. They also make vitamins—probiotics can produce almost all the B vitamins, including niacin, biotin, B6, B12, and folic acid, as well as make one B vitamin into another. They can even make vitamin K, in some circumstances. They protect you against germs: With proper cultures in your small intestine, even salmonella contamination would not affect you and it would be impossible to get a so-called "yeast infection." Probiotics neutralize toxins, preventing their absorption into the body. They have one other key role: controlling unfriendly bacteria and other harmful microforms, preventing overgrowth.

In a healthy, balanced human digestive system, you'd expect to find *three to four pounds* of probiotics. Unfortunately, I estimate that most people have less than 25 percent of the normal amount. Eating animal products and processed foods, ingesting chemicals, including prescription and over-the-counter medicines, overeating, and excess stress of all types disrupt and weaken the probiotic colonies and compromise digestion. That, in turn, allows the overgrowth of symptom-causing microforms and all the problems that come along for the ride.

Probiotics, and the conditions that favor them, discourage symptom-causing microforms and can even cause them to devolve into harmless forms. Here again, pH is critical. The pH inside the small intestine should be basic (7.5–8.0), but within your stomach and colon, being slightly acidic is what you're after. Acidity within a tight range encourages probiotics and prevents growth of harmful yeast, fungus, and other microforms, including parasites. Without the correct pH, probiotics won't function properly. In fact, in a harsh environment, even friendly probiotics may indulge in survival behavior, wreaking the same kind of havoc on your body as their not-so-friendly cohorts. (It is interesting to note that probiotics themselves, though helpful in the digestive tract, are destructive in blood and other tissues.) Furthermore, a mildly acidic environment is required to initiate peristalsis, the rhythmic muscular contractions of the intestinal wall that keep materials moving through.

Acidity in the stomach and colon varies depending on the food you eat. High-water content, low-sugar foods, like those recommended in the program, cause less acidity. High-sugar and high-protein foods increase acidity. As the food moves on into the small intestine, if necessary the pancreas adds alkaline substances (8.0–8.3) to the mix to raise the pH. So the body has ways to moderate the acid or base to appropriate levels. But on our current highly acidic diets, we overtax those systems. Eating right to begin with keeps the stress off the body and lets the process proceed naturally and easily.

A newborn infant already has several different kinds of intestinal microforms. No one knows when they "invade," though some guess it is in the birth canal. However, babies born via Caesarian section have them as well. I believe they don't invade at all—that they are, rather, specialized cells of our body that actually evolve from microzymas already within us. Just as harmful microforms don't have to "infect" us to cause disease symptoms, neither do the helpful ones.

THE SMALL INTESTINE

The twenty-seven feet of small intestine deserve a bit more attention than I gave them in that quick overview. You also need to know that its inner walls are covered with little projections called *villi*, which serve to increase the surface area available, meaning more places for the food to contact on its way through and more capacity to absorb the good stuff from that food. All told, you've got about 7200 square feet or 670 square meters of surface area in your small intestine—about the same as a tennis court! And you need it. Your life literally depends on your body's ability to absorb nutrients from food.

The body's design does not fool around when it comes to getting this job done.

That's why it is such a travesty to interfere the way most of us (unknowingly) do, creating the conditions that allow the explosive growth of yeast and fungus and the toxic effects that come in their wake. Those and other microforms interfere with nutrient absorption. They can cover large sections of the membrane lining the inside of the small intestine, displacing probiotics and preventing your body from getting the good stuff out of what you eat. This can leave you starving for vitamins, minerals, and especially protein, regardless of what you actually put in your mouth. I estimate that more than half of adults in the United States are digesting and absorbing less than half of what they eat.

Overgrowth of microforms, which feed on the nutrients we should be getting (and make their poisonous waste out of them), just make things worse. Without proper nutrition, the body can't heal or regenerate its tissues as necessary. If you cannot digest and assimilate food, the tissues will eventually starve. That not only decimates your energy levels and makes you feel sick, but also accelerates the aging process.

But that's only part of the problem. Consider too that when the villi grab the food passing through, they transform it into red blood cells. These red blood cells circulate throughout the body and transform themselves into body cells of all different types, including heart, liver, and brain cells. I don't think you'll be surprised to learn that the pH of the small intestine must be alkaline in order for the food to be transformed into red blood cells. So, *the quality of the food we eat will determine the quality of the red blood cells that determine the quality of bones, muscles, organs and so on.* You are, quite literally, what you eat.

If the intestinal wall is overgrown and coated with sticky mucus (see page 53), these crucial cells cannot be properly formed. The ones that do get made are underweight. The body must then resort to making red blood cells from its own tissue, stealing from bones and muscles, among other places. Why do body cells transform back into red blood cells? The number of red blood cells must stay above a certain level for the body to function—for us to live. We usually have about five million per cubic milliliter, and the number rarely reaches fewer than three million. Below that, the supply of oxygen (which the red blood cells deliver) won't be enough to support the organs, and eventually they will stop working. Rather than let that happen, body cells begin to revert to red blood cells.

MEET YOUR COLON

The colon, or large intestine, is the sewer plant of the body. It moves out unusable waste while acting like a sponge, squeezing water and mineral content out into the bloodstream. In addition to probiotics like those discussed above, the bowel houses some helpful yeast and fungus that help soften the stool, aiding prompt and thorough elimination of waste.

By the time your digested food hits the colon, most of the fluid material has been extracted. That's as it should be, but it does pose a potential problem: If the final phase of digestion doesn't go just right, the colon can get caked with old (toxic) wastes.

The colon is very sensitive. Any injury, surgery, or other stress, including emotional upset and negative thinking, can change its friendly resident bacteria as well as its general ability to function smoothly and efficiently. Incomplete digestion here sets the stage for intestinal imbalance throughout the digestive tract, and for the colon to become a literal cesspool.

Digestive difficulty throughout the intestines often prevents the proper breakdown of proteins. Partially digested proteins, not usable by the body, can still be absorbed into the blood. In this form, they serve no other purpose than to feed the microforms, increasing the amount of waste they produce. These protein fragments also stimulate an immune system response.

Joy's Story

No one has time to be sick, especially when others depend on you. I'm a single parent also caring for a recently handicapped father, and I need all my strength to manage my home life. But I've been ill for more than two decades. It finally got to the point where I found it easier to stay home and basically remove myself from the human race.

One day at the library, trying to pull myself together after one of my excruciatingly painful attacks, I came upon a book with a chapter about irritable bowel syndrome—a condition I'd been diagnosed with, along with a score of other labels over the years. Its mention of aloe vera and acidophilus sent me immediately to a nearby health food shop, where I started asking questions.

The clerk was quite helpful. She asked why I was seeking these products, and I told her about my IBS— and my thyroid and adrenal dysfunction, hiatus hernia, endometriosis, kidney infections, and numerous other infections. Antibiotics were a way of life for me. My doctors eventually just told me to learn to live with it, but the clerk assured me she knew of people with stories similar to mine who had reversed their conditions. She introduced me to one woman whose story was similar to mine. We clicked immediately, and she told me about how the Youngs' program had changed her life.

I knew beyond a shadow of a doubt what I had to do. I changed my diet immediately and started a regimen of antifungals as well as healthy flora replacement. Within two months I was no longer a prisoner to pain. I feel much better, on my way to 100 percent. An enormous weight has been lifted off my shoulders. My life has gone nowhere but up.

MORE ABOUT MUCUS THAN YOU EVER KNEW YOU WANTED TO KNOW

Although we tend to associate it with head colds and worse, mucus is, in fact, a normal secretion. It is a clear, slippery substance the body makes to protect the surfaces of membranes. One way it does that is by coating anything you ingest, even water. So it also engulfs any toxins you happen to take in, and in doing so it becomes thick, sticky, and cloudy (as we see when we suffer from colds) to "trap" the toxins and escort them out of the body.

Most foods Westerners eat most often cause that thickened mucus. They either contain toxins or break down in a toxic way in the digestive tract (or both). The worst offenders are dairy products, followed by animal protein, white flour, processed foods, chocolate, coffee, and alcoholic beverages. (Vegetables do not cause the formation of this sticky mucus, which is just one more reason to feature them prominently in your diet.) Over time, these foods can encrust the intestines with thick mucus and the fecal material and other debris it traps. This slime is bad enough on its own before you consider that it creates an environment that also promotes the growth of negative microforms.

Emotional stress, environmental pollution, lack of exercise, insufficient digestive enzymes, and the absence of probiotics in the small and large intestine all contribute to the buildup of that slime on the wall of the colon. With buildup, transit time for materials passing through the lower bowel increases. Low levels of fiber in your diet slow it still further. As the gooey mass begins to stick to the wall of the colon, a pocket is formed between the mass and the wall, which is an ideal home for microforms. Material gradually adds itself to the slime, until much of it stops moving altogether. The colon absorbs what fluids are left, the buildup begins to harden, and the home for unfriendly organisms becomes a fortress.

Heartburn, wind, bloating, ulcers, nausea, and gastritis (irritation of the walls of the intestines due to wind and acid) are all a result of a gastrointestinal tract overgrown with microforms. So, too, is constipation, which in addition to being an unpleasant symptom causes more problems and more symptoms. Constipation often shows up as, or comes along with, a coated tongue, diarrhea, cramps, wind, foul odor, intestinal pain, and various forms of inflammation, such as colitis and diverticulitis. (We've all heard the remark that a self-centered person thinks their "stuff" doesn't stink. The solemn truth is, it isn't supposed to! If it does, that's Nature hammering a warning on the door.)

Worse, the microforms can actually bore through the colon wall into the bloodstream. That means not only that the microforms themselves have access to the entire body, but also that they bring their toxins and intestinal matter along with them into the blood. From there they can travel quickly and take hold anywhere in the body, invading cells, tissues, and organs easily enough. All this severely stresses the immune system and the liver, as they desperately try to ward off what doesn't belong. Unchecked, microforms burrow deeper into the tissues and organs, the central nervous system, the skeletal structure, the lymphatic system, and the bone marrow.

This is not simply a matter of clean pipes. This kind of impaction can affect all other parts of the body because it interferes with what should be automatic reflexes and sends inappropriate messages of its own. A reflex is a nerve pathway in which the impulse goes from the point of stimulation to the point of response without going through the brain—as when your doctor taps your knee with that little rubber hammer and your lower leg kicks out. Reflexes can also respond at places other than the one actually being stimulated. Your body is a mass of reflexes. Some key ones are in the lower bowel, connected via nerve pathways to every major organ system in the body. The impacted materials are like a whole squadron of little rubber hammers banging away in there, sending disruptive impulses to other parts of the body. (This is, for example, a major reason for headaches.) That alone can disturb and weaken any and all body systems.

The body creates mucus as a natural defence against acids, as a way to bind them up and get them out of the body. So mucus is not, on the face of it, a bad thing. In fact, it is saving our lives! For example, when you eat dairy, the lactose sugar is fermented to lactic acid, which is then bound to mucus. If not for the mucus, the acid could burn a hole in your cells, tissues, or organs. (If not for the dairy, there wouldn't be a call for that mucus.) It is just that if the diet continues to be excessively acidic, too much mucus is created and the mucus/acid mixture gets sticky and congestive, causing poor digestion, cold hands, cold feet, lightheadedness, nasal congestion, lung congestion (as in asthma), and continual throat clearing.

RESTORING HEALTH

At baseline, we must replenish the probiotics that inhabit our digestive tracts. With proper diet, a normal population will return on its own. You can help move the process along with probiotic supplements.

These supplements have been so highly touted in some corners you might think they are a magic-bullet cure-all. But they won't work all on their own. You can't just throw cultures at the intestines without making the appropriate nutritional changes to maintain proper pH balance, or they'll just pass right through with little effect at all. Or, they could turn on you. You must prepare the environment as well as possible (as laid out later in this book) *before* you take probiotic supplements.

When you do choose a supplement, keep in mind that there are generally different bacteria predominant in the small intestine and colon, since each organ serves a different purpose and has a different environment (acid or base)—for instance, the good bacteria lactobacillus, which requires an alkaline environment, in the small intestine, and bifidobacterium, which thrives in a mildly acidic environment, in the colon.

Neither one will necessarily take once they get to the intestine—and definitely won't if you simply down supplements without making any other changes. Even if they don't, they may still improve the environment on their way through, helping the good bacteria already living there to proliferate. They do have to be able to survive the digestive process, and the best products are designed to help them do that. Bifidobacteria would have an especially long trip if you were to take them orally, as they have to get all the way through the small intestine to the colon. They cannot survive the alkalinity of the small intestine, and so should be administered through the rectum, via an enema. Furthermore, you must take lactobacillus and bifidobacterium separately, as they can cancel each other out if taken together (unless the bifidobacterium is taken rectally).

Another option is prebiotics—special foods that only probiotics eat—which encourage the development of the "friendly" bacteria already in your system. A family of carbohydrates called the fructo-oligosaccharides—FOS—feed bifidobacterium in particular, and lactobacillus as well. They can be taken as a supplement, on their own, or in a formula. You can also get them straight from the source, in asparagus, Jerusalem artichokes, beetroots, onions, garlic, and chicory.

Either way, each individual situation is somewhat different. Be conservative on your own behalf. If you have any doubts that you are getting the right thing, or that it is working as it should for you, consult an experienced health care practitioner.

Besides improving your overall health and helping you lose weight, following this program will clean your intestines and restore your probiotics as well as balance your pH and control the growth of microforms. All of which, you can see by now, intertwine. As the pH of the blood and tissue normalizes, and the intestines are cleansed, nutrient assimilation and waste elimination will too, and you'll be on your way to complete, radiant healthiness.

KATE'S STORY

I went years with a collection of mysterious ailments before a doctor finally diagnosed me as having amoebas and parasites I had picked up in South America. My digestive system was so far gone that it remained out of whack even after I received treatment. Nothing I tried, mainstream or alternative, seemed to restore my normal good health. But with these straightforward dietary changes, and the addition of powdered concentrate of greens (including wheat grass, barley grass, and kamut grass), I'm now able to digest my food, and I've seen tremendous improvement in my overall health and energy. I've got no more abdominal pain, wind, indigestion, or nausea. I am fifty-seven and feel twenty-seven.

I had been on a low-fat, low-sugar diet, and though I wanted to lose weight, I just couldn't cut back on the amount of food I was eating. Whenever I did, fatigue really hit. By just taking out the foods recommended here (for me, that meant mostly getting rid of meat, except moderate amounts of fish, yeast products, dairy foods, refined white flour products, and most fruit)—yet still eating approximately the same number of calories and certainly never going hungry—I lost thirty-five pounds that wouldn't come off for the world before, through traditional diet and exercise schemes.

My husband is a medical doctor, and when he saw my results, he looked into the program for himself—and then changed his diet, too.

Chapter 4

E Is for Eating Right

I realize all that information about microforms and myco-toxins and mucus can be a bit overwhelming—not to mention unappetizing! I do want you to get just a bit of the science behind our recommendations, so you really understand how you can best serve your own body. I also want you to understand the urgency of it. But don't worry—the solutions are a lot simpler than the problems. And they are well within your reach.

Part II of this book lays out the specifics of exactly how to put this plan into practice in your everyday life. The recipes in Part III will give you a great sendoff down this new path. Right here and now, I'm going to give you the most basic outlines of this diet: the foods to include, and the foods you must avoid.

The idea behind the whole thing is to keep your body basic, thereby taming the microforms, eliminating mycotox-ins, and ensuring radiant health. To that end, you're going to focus on foods that alkalize your body and avoid those that acidify it.

All food digested in our bodies metabolizes, or burns, down to an ash residue. This ash residue can be neutral, acidic, or alkaline, depending mostly on the mineral content of the original food. For example, potassium, calcium, magnesium, sodium, zinc, silver, copper, and iron form basic ash; sulphur, phosphorus, chlorine, and iodine leave acid ash. Most elements are alkaline.

Fortunately for us, it is easy to categorize which foods leave which kind of ash. In general, animal foods—meat, eggs, dairy—processed and refined foods, yeast products, fermented foods, grains, artificial sweeteners, fruit, and sugars are acidifying, as are alcohol, coffee, chocolate, black tea, and fizzy drinks. Vegetables, on the other hand, are alkalizing. That includes a few that are technically fruits: avocado, tomato, bell pepper (and mushrooms are an exception). A few nonsweet citrus fruits are also basic in the body, as are sprouted seeds, nuts, and grains. Grains are acidifying, though a few (millet, buckwheat, and spelt) are only very mildly so. Raw foods are more alkalizing, while cooked food is more acidifying.

To maintain a balanced pH in your blood and tissues, your diet should consist of at least 70 to 80 percent basic foods—that is, no more than 20 to 30 percent acidifying foods (and at least half of that 70 to 80 percent should be raw). These are *visual* measurements, not by weight or calories. The more alkaline your diet, the more rapid your improvement will be. Unlike the familiar "food pyramid," which has an overall acidic effect, this program will bring you back to basic.

The initial phases of the program (as described in Part II) are the most restrictive. The guidelines here are essentially for maintenance, once you've gotten your body on track. These general principles I have divided into two sections: what you should eat, and what you should avoid, with all the whys and wherefores. First, we'll look at what you must include in your diet: the good alkalizing foods, including vegetables, non-animal protein, raw foods, and pure water. Then we'll dive into the acidifying foods you must avoid, such as meat, sugar, and yeast.

WHAT TO EAT

Vegetables

Vegetables should be the focus of your diet, the majority of any meal you sit down to, the substance covering most of your plate. However you want to look at it, these vegetables should be your new best friends. They are some of the lowest-calorie, lowest-sugar, most nutrient-rich foods on the planet. Beyond that, they provide (as detailed below) vitamins and minerals, fiber, chlorophyll, enzymes, phytonutrients, and alkaline salts that control microforms and their mycotoxins. Who could ask for anything more?

Choose from a wide variety of vegetables, making the majority of them green. Here's why: Vegetables are an excellent source of the **alkaline salts** that protect against microform overgrowth and the associated mycotoxins as well as help neutralize acids in the blood and tissues. So basically, the more, the merrier. Fresh is essential, and organically grown is preferable.

Vegetables' green color is produced by **chlorophyll.** I call it the blood of plants, as its molecular structure and chemical components are similar to that of human blood. Blood's hemoglobin is made up of carbon, hydrogen, oxygen, and nitrogen organized around a single atom of iron. Chlorophyll is the same with the exception of the central atom, which is magnesium rather than iron.

Chlorophyll helps the blood cells deliver oxygen throughout the body. It also reduces the binding of carcinogens to DNA in the liver and other organs. If that is not enough of a benefit for you, keep in mind that it also breaks down calcium stones—stones that the body creates to neutralize and dispose of excess acid for elimination.

Green vegetables and particularly leafy greens have the highest amounts of chlorophyll.

Vegetables, and particularly green vegetables, are incredibly nutrient-dense and provide just about all the **vitamins, minerals, and micronutrients** you could ever need.

Vegetables are loaded with the **fiber** that is crucial to your diet. Besides the accepted benefits of fiber in reducing cancer and other serious health concerns, studies have shown that fiber markedly decreases mycotoxicity. Fiber acts like a sponge, soaking up acids from the body. It also works like a broom, cleaning out the intestines.

JULIA'S STORY

Doctor after doctor said I could never conceive without medical intervention because of an immunological disorder. After ten years of fertility treatments, my body was full of fertility drugs and needle scars. Now that the treatments were over with, I was a great candidate for cancer. There was a solution for that problem too: birth control pills and further hormonal roller coaster rides. Worst of all: still no baby.

Then we learned about The pH Miracle program. I felt I was blessed with a second chance, and started "cleaning house"—in my own body, using sound nutrition to undo all the mistakes I had made in my twenties and early thirties. The knowledge of how my body truly works gave me the power to control the way I looked and felt, which brought with it inner peace and confidence. More than the physical effect, I felt a spiritual change come over me. It seemed as though my once depressed, tired, and preoccupied self became one hopeful, energetic, and peaceful entity. Internal dueling was replaced by a sense of clear purpose.

Several months after I started on this program, I became slightly ill and discovered, to my complete surprise and joy, that I was pregnant! It came as an enormous shock to me, considering all the difficulty I had in the past, to become pregnant without even trying. Our bouncing "Green" baby was born last year, our own miracle child. She is happy, healthy, and smart and beautiful! After six months of breast feeding, she started on solid foods, which I prepare myself using almond milk, organic grains, and veggies (broccoli, cabbage, carrots, celery, green beans, spinach, parsley, fresh coriander, dill), avocado, new potatoes, and yams. She loves munching on cucumbers and peppers. She loves pureed salads with lemon juice and olive oil. I know our happy ending would never have been possible without our new way of life.

Vegetables are plentiful sources of the **enzymes** that are needed for just about every chemical activity in the human body. There are thousands of them, and we need them all for our overall energy reserves. The faster we exhaust our supplies, the faster we die, basically.

Among many other things, enzymes aid in digestion. Different enzymes are called upon according to the foods eaten, depending on the type of food and how it was processed. Whatever the enzymes within the food can accomplish in the way of digestion makes less work for the digestive enzymes of the body. The conservation of all that potential, then, provides a boost in your overall energy. The body can make more metabolic enzymes, instead, for use by organs and tissues carrying on their metabolic functions, and to provide repairs associated with mycotoxin damage.

Heat (over 118°F or 50°C) destroys enzymes, which is why you want to get a lot of your food uncooked—or at least cooked as little time as possible. Cooked food requires the body to produce all the necessary enzymes, creating an unnecessary stress and diverting resources from other jobs.

CHEW THIS OVER

Besides not destroying them, the best thing you can do to help enzymes do their work is to chew your food extremely well. The idea is to drink your solids and chew your liquids. That means chewing solid food long enough to convert it to a liquid consistency in the mouth—and not gulping liquids without mixing them a little with your oral secretions. Although the latter process may not be featured in the etiquette books, it gives the digestive process the boost it needs to extract all the value from food it can.

You should also be sure to take small bites. When solid food is properly chewed, it expands considerably. Large initial mouthfuls then provoke premature swallowing. Finally, stop eating before you think you are full, and give your stomach a chance to register it has had enough.

Phytonutrients, or phytochemicals as they are also known, are highly biologically active and extremely beneficial. They exist in astounding number and variety. Some phytochemicals give plants their color, such as the yellow, orange, and red in summer squash, carrots, and peppers (though it is chlorophyll that takes care of the green, of course).

Phytonutrients help prevent cancer, lower cholesterol, relieve arthritis and osteoporosis, stop hormones from being turned into acids, and more. Some argue that this is because they counter free radicals, but in truth, although the effects have been observed, the pathway is rarely understood. I believe their secret is, rather, their ability to eliminate harmful microforms and their toxins. Phytonutrients generally bind to (thereby neutralizing or eliminating) acids.

GET YOUR FRESH VEGGIES HERE!

Here is a partial list, in alphabetical order, of excellent dark green and yellow vegetables. Eat freely:

Asparagus
Aubergine

Beetroots
Broccoli
Brussels sprouts
Burdock
Cabbage
Carrots
Cauliflower
Celery
Cucumbers
Garlic
Green and yellow squash (courgette and summer squash)
Green beans
Greens of all kinds (including spinach, mustard greens, collards, kale, lettuce, watercress, and Swiss chard)
Okra
Onions
Parsley
Parsnips
Peas (fresh)
Radishes
Red and yellow and green peppers
Salsify
Sea vegetables such as nori, wakame, and hijiki
Spring onions
Sprouted grains or beans or seeds
Swedes
Turnips
Water chestnuts
Winter squash
(For as long as you are having symptoms, go easy on the

> high-sugar vegetables such as carrots, beetroots, and winter squash.)

One large group of phytonutrients are the bioflavonoids, water-soluble companions of vitamin C that abound in the plant kingdom.

Grasses

What I said about the benefits of vegetables goes double for grasses.

Right away I'm sure you noticed a potential problem here. You are not, in fact, a cow. What are you supposed to do—save the clippings when you mow the lawn?

Not to worry. While you can get wheat grass in most health food shops, and dishes flavored with lemongrass at Thai restaurants, that's pretty much it for the grasses generally sold for human consumption. But what we recommend is using a supplement to get your grasses—either taking capsules of powder or, preferably, mixing the powder into water for a "green drink." I'll get more into the details in later chapters, as all "greens" supplements are not created equal. Basically, you want to look for a wide variety of grasses (such as wheat grass, barley grass, oat grass, dog grass, kamut grass, lemongrass, and shave grass/horsetail), while avoiding all algaes, mushrooms, and such (see "What to Avoid" on page 87 for details).

Grasses are incredibly nutrient-dense, even more so than vegetables generally. (After all, how do you think a cow gets by?) Wheat grass and barley grass are particularly good sources of chlorophyll, for example, and it is the chlorophyll that gives grasses the power to regenerate our bodies at the molecular and cellular level. To give you just two examples:

- Wheat grass contains more than one hundred food elements, including every identified mineral and trace mineral and every vitamin in the B-complex family. It has one of the highest pro–vitamin A contents of any food, and is rich in vitamins C, E, and K. Wheat grass juice is 25 percent protein, a higher percentage than in meat, fish, eggs, dairy products, or beans. In addition, it has high amounts of an antifungal, antimycotoxic substance called laetrile.
- Barley grass boasts four times as much thiamine (vitamin B1) as whole wheat flour, and thirty times as much as milk. It has even more vitamin C than oranges (actually, seven times as much vitamin C).

Low-Carbohydrate Vegetables

Complex carbohydrates make a lot of acid when they break down, and so should not exceed 20 percent of your diet. So the vegetables you choose should be mainly those that are low in carbohydrates (like the fresh veggies listed in the box on page 66), and you should enjoy legumes and grains in limited quantities because they are high in complex carbohydrates.

Diets with 50 percent or more complex carbohydrates—which would encompass a lot of seemingly "perfect" low-fat, vegetarian diets as well as the official food pyramid—provide a favorable environment for microform overgrowth in even healthy digestive tracts. Simple carbohydrates—fruits, sugars, white flour—are even more conducive to overgrowth, and the typical Western diet is full of them.

High-carbohydrate vegetables, including potatoes, winter

squash (acorn, butternut, pumpkin), yams, and sweet potatoes can be eaten in moderation. When you do eat them, make sure they are fresh, not stored for a long period of time, and check them carefully for fungal spots (especially potatoes). Red new potatoes are the best choice in the potato family because they are this year's potatoes. Some potatoes could be *very* old by the time you get them, having been stored in silos. Red potatoes are fresher than other potatoes (except those you pick from your own garden), but still should be used sparingly.

If we lived in an ideal world, you'd be able to choose fresh legumes (beans and peas) as well. But almost all legumes sold in the shops are dried or canned—though I do get frozen edamame (soybeans) in my local health food shop. Avoid the canned beans, by all means, but if you pick over, soak, and rinse the dried beans thoroughly before cooking, you can use them in that 20 percent of your diet. Beans are actually quite antifungal—even if you find a few bad beans, you never see the whole bag get moldy, so in moderation they are a good choice. Their starchiness is the main reason to limit quantities. The best choice of all is to eat *sprouted* beans (see page 82).

Most legumes are primarily starch: kidney beans, pinto beans, adzuki beans, black beans, string beans (navy or white beans), chickpeas (garbanzo beans), split peas, black-eye peas (cowpeas), and lima beans. Two are primarily protein and so are okay to include more often: soybeans (especially edamame, the fresh, whole bean) and lentils.

The grains you eat must also be fresh, not stored. Stored grains are any grains stored into the next season. Look for a supplier that can assure you you're getting this season's grains—grains not more than three months away from having been harvested. Ask at your shop to ask their supplier. Stored grains are full of fungus and their mycotoxins (see

"What to Avoid" on page 87 for details). Grains form acids when they are broken down, so limit how much you eat of them, and use only fresh, organic grains. Sprouted (see page 82) is best.

The most common complex-carbohydrate grains are the most acid-forming: wheat and rice. (They make mucus, too.) They have to stay in that 20 percent of your diet. Corn you should avoid altogether (more about that coming up). On the other hand, amaranth, quinoa, and spelt are only slightly acidic, and millet and buckwheat hover between neutral, slightly acid, and slightly alkaline, so they don't contribute to the formation of sticky mucus.

There are other benefits to these grains as well. Buckwheat groats (really a seed, strictly speaking, but used like a grain) and millet are high in protein and digest slowly, keeping blood sugar balanced. Spelt contains more protein, healthy fats, and fiber than wheat. Spelt is also plentiful in mucopolysaccharides—vital complex sugars that literally glue the body together, lubricate joints, and support immune function. It is also high in B17, which is an anti-cancer vitamin.

Raw Foods

Because cooking destroys those all-important enzymes, the more of your vegetables you eat raw, the better. Aim to have *at least* 40 percent (visually) of your food uncooked—working up to 75 to 80 percent as you get used to this program. Think salads. Great, big salads. In infinite variety.

Raw foods also contain energy or life force (technically, they are "biogenic"), which they can transfer to you, while cooked foods are dead.

Fresh, Organic Foods

Organic foods let you avoid exposure to pesticides and all the other chemicals routinely dumped on most produce. They are usually more nutritious, as well, since the soil (where plants draw their nutrients from) is less depleted than it becomes with the harsh treatment on standard farms. Organic produce is as much as 300 percent higher in nutrients than non-organic.

Eating your food as fresh as possible is also key. The minute something is picked, the nutrients begin to break down. Ideally, we'd still be living in the kind of world where we all could walk out into our gardens, pluck our dinner, prepare it immediately, and sit down to enjoy it within the hour. I realize that's not feasible for most of us, most of the time, but to get the most out of what you eat, look to get as close to that experience as you can.

When organic is not a choice, get the freshest produce you can and clean thoroughly by soaking them for 10 to 15 minutes in water with chlorite, ClO_2 (or hydrogen peroxide—rinse well afterward), added (60 drops to the gallon).

Water

Chapter 5 looks at water in depth, but the main message bears repeating: Plenty of water—correction, clean, *pure* water—is absolutely crucial to creating a healthy pH and a healthy body. The body is 70 percent water—and the blood even more so: 94 percent water—and we should provide it with a lot of this basic component. That includes foods that are actually high in water. That means—you guessed it— vegetables!

Non-animal Protein

Here's a question we get a lot: "Where do you get your protein?" The inquiry presupposes, first, that protein comes only from meat, dairy, and eggs, and second, that getting enough protein is somehow difficult. The truth is there is plenty of protein in plants. And if you are getting enough calories to be healthy, and you are eating a reasonable variety of foods, you're getting enough protein. Don't just take my word for it: a clinical study published in the very mainstream *Journal of the American Dietetic Association* analyzed the diets of meat-eaters, vegetarians who eat dairy and eggs, and pure vegan vegetarians using strict requirements about how much protein would easily cover the requirements of growing children and pregnant women. Not only did all three diets provide enough, they *all* actually doubled requirements. The take-home message is, no one has to worry about getting enough protein. If you eat a normal amount of reasonable foods, you will.

Most people seem to think protein needs to come from meat and dairy products. Even those more "in the know" about alternative health subscribed for a long time to the theory that vegetable proteins were somehow second-class and required proper "combining" to be complete. But vegetables carry all the amino acids (the building blocks of protein) the body needs. Not every vegetable has every one, of course, but if you are eating a wide variety of vegetables, especially dark green and dark green leafy vegetables, and supplementing with grasses, you are getting plenty of all the essential amino acids.

The body has a free amino acid pool, which contributes about seventy grams of protein daily. We all have these protein reserves, so unless you have specific symptoms of protein deficiency (muscle tissue loss, hair falling out, brittle nails), you can be sure you are getting enough protein.

We eat fish maybe twice a month. Other than that, we eat tofu a couple times a week. We often eat legumes, and get plenty of raw nuts, sprouted seeds and grains, and avocados, all of which have high-quality protein that is better assimilated than animal proteins are. All the rest of the protein we need we get from greens. The key to providing your body with protein is quality, *not* quantity.

PERCENTAGE OF CALORIES FROM
PROTEIN IN ALKALIZING FOODS

Food	Protein Calories	Food	Protein Calories
Vegetables		Chives	18%
Alfalfa sprouts	40%	Collards (leaves)	48%
Artichoke	29%	Collards (stems)	36%
Asparagus	25%	Courgette	26%
Aubergine	12%	Cress	26%
Bamboo shoots	26%	Cucumber	10%
Beet greens	22%	Dandelion greens	27%
Broccoli	49%	Fennel	28%
Brussels sprouts	49%	Garlic	20%
Cabbage, Chinese	12%	Kale (leaves)	60%
Cabbage, red	20%	Leek	22%
Cauliflower	27%	Lettuce, Boston	12%
Celery	10%	Lettuce, green-leaf	42%
Chard, Swiss	24%		

Food	Protein Calories	Food	Protein Calories
Lettuce, iceberg	27%	**Legumes**	
Lettuce, loose-leaf	13%	Chickpea	25%
Mustard greens	22%	Lentil	30%
Okra	24%	Lima bean, fresh	9%
Onion (green)	15%	Mung sprouts	38%
Parsley	36%	Pea, green fresh	6%
Pepper, green	12%	Red bean, dried	23%
Pepper, red	14%	Soybean, dried	34%
Pepper, red hot	13%	Soybean, fresh	11%
Radish	10%	Soybean sprouts	6%
Rhubarb	11%	String (white or navy) bean	26%
Seaweed, dulse	25%	Tofu	43%
Spinach	49%		
Turnip greens	30%	**Nuts and Seeds**	
Watercress	22%	Almond	19%
Wheat grass	25%	Brazil nut	14%
		Hazelnut	13%
Fruits		Pumpkin seed	29%
Avocado (Haas)	22%	Sesame seed	19%
Avocado (Ryan)	15%	Sunflower seed	24%
Grapefruit, sour	5%	Sunflower seeds, sprouted	33%
Lemon	13%		
Lemon juice	5%	**Grains**	
Tomato, green	12%	Barley	10%
Tomato, red	18%	Millet	10%
		Rice, brown	8%
		Wheat	17%
		Wheat bran	16%

Soy

Soy is a smart addition to your diet, as a source of protein and a wide variety of nutrients. Soybeans contain a host of beneficial chemicals:

- Isoflavones, a type of phytoestrogen ("plant estrogen"), help prevent the growth of hormone-dependent cancers, such as many breast cancers.
- Daidzein, a particular isoflavone, inhibits the growth of cancer. It also promotes cell differentiation in animals—cancer cells being undifferentiated.
- Genistein is an enzyme that can inhibit tumor growth and promote cell differentiation. Studies have shown that it helps block the growth of prostate cancer cells and breast cancer.
- Protease inhibitors (more enzymes) block the action of enzymes that may promote tumor growth and work against a wide range of cancers, including some of the most common—colon, breast, and liver cancers.
- Physic acid chelates (combines with) mycotoxins that promote tumor growth, binding to them and taking them out of the body. Studies have shown that it can help reduce the size and number of tumors in laboratory animals fed mycotoxins.
- Saponins, studies have shown, lower the risk of certain cancers, including those of the breast, prostate, stomach, and lung. (Saponins are also found in chickpeas and ginseng.)
- Soy works to help prevent some cancers, protect the heart, and balance the hormones.

Your best bet is fresh soy sprouts (see "Sprouts" on page 82) or a soy sprouts supplement (see Chapter 10). Other choices are soybeans, edamame, tofu, soybean oil, and lecithin (a soy by-product that comes as liquid or granules to be used in recipes or sprinkled on salads and soups, or taken as a supplement). Soy milk, however, is usually *not* a good option, as almost all kinds contain added rice syrup to sweeten them—and the syrup ferments and creates more acidity. Nonsweet soy milk is okay. Organic is best, as always, and fresh is crucial.

I recommend sprouts without reservation, but whole beans are slightly acidic, and processed soybeans require some cautions. In processing soybeans, the intact bean is altered—cells broken, protective coating removed—which activates the inherent microzymas to start decomposing their surroundings. And of course, many soy products (such as soy hot dogs) are just as refined and processed as any junk food. I wish we all had access to fresh tofu whenever we wanted it, so we could eat it on the same day it was produced. But in this reality, your best bet is simply to use a packaged brand that is dated, rinse it thoroughly, and use it as soon as possible. Don't let it sit in the refrigerator for long.

With that in mind, tofu is a very good source of protein, and is certainly far better for you than animal and dairy foods. (One study pinpointed eating tofu as the single factor associated with a lower risk of developing prostate cancer among Japanese men living in Hawaii. Another found a strong correlation between consumption of soy foods and a decreased risk of developing breast cancer in Chinese women.) Tofu is a good food for making the transition to a vegetarian diet, and occasionally as part of a balanced, wholesome diet.

On the other hand, you must avoid fermented soy products such as miso, tempeh, and soy sauce (see page 104).

Fresh Fish (Occasionally, or as a Transition Food)

The water fish live in is full of fungus, and they couldn't survive if they weren't resistant to it. And fish is rich in healthy omega-3 oils (essential fatty acids), protein, and several nutrients. Still, it shares the properties of all animal foods: It has no fiber and forms acid and sticky mucus. On balance, fish is still a good choice for making the transition to a vegetarian diet, or used on occasion in the context of a healthy (basic) diet.

If you do choose to eat fish, it must be absolutely fresh. If it is not newly caught, or you can smell that fishy smell, it is already spoiling so you must avoid it. You want to make sure it comes from unpolluted water. Choose salmon, trout, red snapper, swordfish, and tuna, for their high levels of omega-3 oils, but be sure to avoid shellfish altogether. Shellfish are scavengers—they eat anything and everything, including the faeces of other fish. As a result, they are full of toxins. Skip dried fish, which is used in many Asian dishes, especially soups, as it has fungus and mycotoxins on it. (Although the US Food and Drug Administration recently announced that swordfish and tuna have potentially high mercury levels and so should be restricted or avoided, at least by women of child-bearing age, I believe the good oils you get from these fish far offset any risk for mercury poisoning. Besides, following the program in this book, the amount of fish you eat *is* limited.)

GREEN ATHLETES

We exercise a lot—running, lifting weights, and walking and hiking every day, and we rely on this program to sustain us. Alkalizing increases athletic performance, building stamina and muscle as well as providing pure energy. So while The pH Miracle can be a literal life-saver for sick people, it is also a boon to even the fittest and healthiest among us. It will make anyone well, but it will also keep you from getting sick in the first place and keep you functioning at peak levels, physically.

We've seen the results in ourselves, of course, but on a more objective level we've seen the results in all kinds of stellar athletes. We advise two biathletes in training for the 2002 Winter Olympics, one of whom attributes his recent third-place finish in a European competition (the Europeans dominate the sport) to the concentrated greens he takes. When he sought us out, he was having trouble even completing some events! We also work with the top soccer team in Trinidad, which uses concentrated green powder to fuel them through all their top-flight international matches. We've also started the West Point gymnastics team on the program (more about that in a moment).

The one that really blows our minds is Stu Mittleman. In summer 2000, Stu ran from San Diego to New York City in fifty-six days (running approximately *two marathons* every day). And he did it all powered by alkaline foods, appropriate supplements, and daily "green drink" (actually, several daily green drinks). He also holds the world's record for long-distance running: If you want to challenge him, you'll have to do better than one thousand miles in thirteen days, or approximately *three* marathons every day.

He does eat fish almost every other day, and he also has a steak now and again. But for the most part Stu is a living, breathing pH Miracle. Vegetables are by far the largest portion of his diet.

You don't have to be at that kind of extreme to experience how The pH Miracle program can enhance your athletic abilities, though we find it inspiring to know what it can do even really "out there." It is important to note it is a two-sided coin: Exercise itself accelerates the lymphatic process (that is, sweating), getting rid of excess acids and wastes through that all-important "third kidney"—the skin. So the program is good for exercising; exercising is good for the program.

Now, about those gymnasts. Here we've got something more than just anecdotal evidence. No matter how impressive Stu is, he's still just one guy, doing his thing. But at West Point I (Rob) was able to do a controlled study. I divided the eleven Army college gymnasts into two groups (A and B) and gave one of them (A) concentrated green powder and pH drops to add to their water every day. (I also had them each wear a special pendant designed to mitigate the negative effects of electrical magnetic frequencies, or EMFs.) Group B had just plain water (and a placebo pendant). Their diets stayed the same, and they made no other changes to their lifestyle.

Over the course of the season, their coach kept statistical results on each gymnast for every meet, on all six events (floor exercise, vault, high bars, parallel bars, pommel horse, and rings), including the number of routines attempted and the number of routines "hit" (successfully completed). The five competitive gymnasts taking the supplements, group A, hit 66 percent of all their routines. The other six, group B, hit 38 percent. Group A outperformed group B by almost 100 percent! And this was just with the addition of green drink and pH drops and the pendant—not even the full pH Miracle program.

In addition, although this is more subjective, I think it is still worth noting that group A gymnasts reported increases in strength and endurance, longer practice time, improved attitudes toward workouts, and most important of all, *less* time in recovery after workouts and meets compared to their group B teammates and to themselves at the start of the season.

I also took before-and-after blood samples of all the gymnasts and found that the blood of group A participants was significantly healthier and stronger than that of men in group B.

Essential Fats

One of the most dangerous fad diets today is the "no-fat" obsession. Fats play a crucial role in our bodies and getting none at all opens our bodies up to nutritional deficiency and the degeneration that comes with it. The key thing is to get *healthy* fats, not the artery-clogging, zero-nutrient varieties most Westerners eat (primarily saturated fats and "partially hydrogenated" oils—liquid oils chemically altered into solids). Approximately 20 percent of your calories should come from these healthy fats.

What your body needs are the *essential fatty acids*. EFAs are, well, essential—vital to good health. They are the building blocks of the fats that strengthen cell walls. Polyunsaturated fats such as linseed (flax), borage, evening primrose, grape seed, and hemp oils help construct cell membranes, produce hormones, and bind and eliminate acids. Most oils contain both monounsaturated and polyunsaturated fats, and those that are predominately monounsaturated, such as olive oil (as well as raw nuts and avocados) are also beneficial. They are used for cellular energy—meaning our body runs on those instead of sugars when we finally get in balance.

EFAs strengthen immune cells, lubricate joints, insulate the body against heat loss, provide energy, and are used to make the hormone-like prostaglandins that protect against heart disease, stroke, high blood pressure, arteriosclerosis, and blood clots, and are necessary for energy metabolism and immune system health. EFAs can also help relieve arthritis, asthma, PMS, allergies, skin conditions, and some behavioral disorders, as well as improve brain function.

Nuts, seeds, and avocados are all good sources of healthy fats, including the omega-3s and omega-6s you may have heard about (sometimes called "fish oils" because fish is another good source if you are not vegetarian). EFAs are found in the highest concentration in linseed (flax seed) oil, borage seed oil, and hemp seed oil. (We like the brands Essential Balance from Arrowhead Mills, sometimes also sold as Omega Nutrition, and Udo's Choice®. You might try those or similar combination oils – see Resources for UK suppliers.) Look for *cold-pressed* oils, extracted and packaged without being heated. Then, don't heat the oil yourself! Add to vegetables after warming or steaming, or make a flavorful salad dressing simply by combining with lemon juice and seasoning. Beware of rancid oils. Buy only what you'll use up pretty quickly, store it in dark containers in a dark place, and only use what smells fresh.

Sprouts

Full of vitamins, minerals, and complete proteins, sprouts are just about the best food you can eat. They are living plant foods that are biogenic—meaning they can transfer their life energy to you! Seeds become more alkaline as they sprout, and sprouts are packed with enzymes.

In the sprouting process, plant hormones are activated, proteins are predigested into easily assimilated amino acids that work better in the human body, fats are broken down into more easily assimilated fatty acids, and starches are broken down into easily assimilated vegetable sugars. The body does need *some* sugar—just not all the hard sugars that are wiping out the pancreas and sapping energy supplies!

Sprouts—and soaked nuts and seeds—are alkalizing, life-generating, revitalizing, high-energy foods. They are high in enzymes, predigested complete proteins, chelated (protein-coated) minerals, nucleic acids, vitamins, RNA, DNA, and vitamin B12. Their plant hormones are activated, their starches broken down into easily assimilated vegetable sugars, their proteins predigested into easily assimilated free amino acids, and their fats broken down into soluble fatty acids. And their nutrient content skyrockets: Biotin content increases by 50 percent at sprouting, vitamin B5 by 200 percent, B6 by 500 percent, folic acid by 600 percent, and riboflavin (B2) by 1,300 percent!

You may think of sprouts only as the familiar bean sprout and alfalfa sprout. But sprouts from just about any beans, grains, or seeds are healthful and delicious. To give you some examples, we enjoy mung bean sprouts, chickpea sprouts, green lentil sprouts, sesame sprouts, sunflower sprouts, buckwheat sprouts, and wheat sprouts. As mentioned, sprouted is the best way to eat soy. In fact, it is the ideal way to eat all legumes: When sprouted, they are much easier to digest and will not produce intestinal wind as they do when full-grown and cooked.

While sprouts are available in health food shops and supermarkets, growing your own ensures absolute freshness and maximum living energy. Sprouts are easy to grow in your own kitchen, in any season, providing you with fresh organic produce year round. (More about how to do this in Chapter 11.)

Herbs and Spices

Herbs and spices provide both flavor and nutrition. Herbal teas can also be very beneficial. Freshness is again key: Herbs and spices can get moldy during processing (drying) and storage.

Juices

Although you lose the benefits of fiber, juicing enhances all the other benefits of vegetables and grasses. (Fruit juice, however, you should avoid. Details follow later in this chapter.) When you "drink your vegetables," your body is receiving a greater concentration of rapidly usable alkaline salts, vitamins, minerals, chlorophyll, and enzymes, and can assimilate them more easily and rapidly.

Juice from green vegetables is highly alkalizing. If you are having any symptoms, you should limit the amount of beet and carrot juice you use, as they are as sweet as they are because they are relatively high in sugar (11 percent and 13 percent sugar, respectively). Carrots are generally alkalizing, but concentrated the way they are in juice, they can quickly add up to too much sugar.

Though not technically juices, I also want to mention vegetable broths here, since they, too, are so alkalizing—particularly cucumber, onion, and garlic. You can make your own or buy premade (just check to make sure it is preservative-free and contains no yeast).

Lemons, Limes, and Grapefruit

Although you must avoid almost all fruits (see details later in the chapter), lemons, limes, and (nonsweet) grapefruit are actually beneficial. (Usually, white grapefruit is less sweet than pink, though taste is your best indicator on how sweet it is.) Fruit itself is not unhealthy *per se*—in fact, most fruit is rich in nutrients—but the sugar it contains ferments like any other sugar, wreaking the familiar havoc in your system. It is crucial to strictly avoid fruit when you are embarking on this program, though once you are thoroughly back in balance, your body will be able to tolerate a piece of fresh fruit, in season, once in a while for a treat.

Though lemons, limes, and grapefruits are chemically acid, my tests show that when they are metabolized in the body they actually have an alkalizing effect. They have very little sugar (lemon and lime 3 percent, nonsweet grapefruit 5 percent). And they contain an abundance of oxygen. Remember, microforms do not live well in the presence of oxygen, so these foods prevent microform overgrowth.

Squeeze fresh lemon or lime into your purified water throughout the day—and especially before going to sleep for the night. Don't take lemon or lime within half an hour before a meal, or for ten minutes after you finish eating. It is best to drink an hour before eating, then wait until digestion has taken place to drink more.

Tomato and Avocado

Tomato and avocado are also good vegetable choices (though technically they are fruits), because eaten raw, they are alkalizing. (When tomato is cooked it becomes mildly acid-forming.) Both are also low in sugar (avocado 2 percent, tomato 3 percent). Avocados are rich in healthy monounsaturated fats, and are a good source of protein, making them a great source of fuel. And they have more potassium than bananas—and *way* less sugar.

THE pH OF FOOD

The following is a list of common foods with an approximate, relative potential of acidity (−) or alkalinity (+), as present in one ounce of food.

Vegetables and Low-Sugar Fruits			
Peas, ripe	+0.5	Green cabbage, December harvest	+4.0
Asparagus	+1.1	Savoy cabbage	+4.5
Artichokes	+1.3	Lamb's lettuce	+4.8
Comfrey	+1.5	Peas, fresh	+5.1
Green cabbage, March harvest	+2.0	Kohlrabi	+5.1
Lettuce	+2.2	Courgette	+5.7
Onion	+3.0	Red cabbage	+6.3
Cauliflower	+3.1	Rhubarb stalks	+6.3
White radish (spring)	+3.1	Horseradish	+6.8
Swede	+3.1	Leeks (bulbs)	+7.2
White cabbage	+3.3	Watercress	+7.7
		Spinach, March harvest	+8.0
		Turnip	+8.0

Lime	+8.2	**Nonstored Organic Grains**	
Chives	+8.3	**and Legumes**	
Carrot	+9.5	Brown rice	−12.5
Lemon	+9.9	Wheat	−10.1
French cut beans		Buckwheat groats	−0.5
(green beans)	+11.2	Millet	−0.5
Fresh red beetroot	+11.3	Spelt	−0.5
Sorrel	+11.5	Lentils	+0.6
Spinach		Soy flour	+2.5
(other than March)	+13.1	Tofu	+3.2
Garlic	+13.2	Lima beans	+12.0
Celery	+13.3	Soybeans, fresh	+12.0
Tomato	+13.6	String beans	
Cabbage lettuce, fresh	+14.1	(navy or white beans)	+12.1
Endive, fresh	+14.5	Granulated soy (cooked,	
Avocado	+15.6	ground soybeans)	+12.8
Red radish	+16.7	Soy nuts (soaked soybeans,	
Cayenne pepper	+18.8	then air dried)	+26.5
Straw grass	+21.4	Soy lecithin, pure	+38.0
Shave grass (horsetail)	+21.7		
Dog grass	+22.6	**Nuts and Seeds**	
Dandelion	+22.7	Wheat kernel	−11.4
Kamut grass	+27.6	Walnuts	−8.0
Barley grass	+28.7	Pumpkin seeds	−5.6
Soy sprouts	+29.5	Sunflower seeds	−5.4
Sprouted radish seeds	+28.4	Macadamia nuts	−3.2
Sprouted chia seeds	+28.5	Hazelnuts	−2.0
Alfalfa grass	+29.3	Linseeds (flax seeds)	−1.3
Cucumber, fresh	+31.5	Brazil nuts	−0.5
Wheat grass	+33.8	Sesame seeds	+0.5
Summer black radish	+39.4	Cumin seeds	+1.1

Fennel seeds	+1.3	Linseed (flax seed oil)	+3.5
Caraway seeds	+2.3	Evening primrose oil	+4.1
Almond	+3.6	Marine lipids	+4.7
Fats (Fresh, Cold-Pressed Oils)		**Water**	
		Distilled water	(neutral)
Sunflower oil	−6.7	Fresh coconut water	+9.04
Ghee	−1.6		
Coconut milk	−1.5	**Fish**	
Olive oil	+1.0	Freshwater fish	−11.8
Borage oil	+3.2		

WHAT TO AVOID

Sugar

Sugar feeds negative microforms like petrol feeds a fire. Microforms love all forms of sugar (white sugar, brown sugar, processed beet, cane, and corn sugars and syrups, maple syrup, honey, molasses, sucrose, fructose, maltose, lactose, glucose, mannitol, sorbitol, galactose, monosaccharides, date sugar, partially refined sugar, sweets, soft drinks, pastries, ice cream, chocolate, carob, and yes, even "natural" sugars from fruit), especially those that cause a rapid rise of blood sugar (cane sugars and corn sugars). Any bread maker knows what happens to the yeast when you add sugar to the dough—the yeast ferments the sugars, causing the bread to rise.

The more sugar harmful microforms get, the faster they will reproduce, and the faster they reproduce, the more they are decomposing and fermenting your body from the inside. Cut them off.

As mentioned, the body does need *some* sugars, but these harsh ones overly tax the body. Better are the gentle vegetable sugars that our organs are much more able to handle. You'll get plenty of what you need in that regard following this program.

Be sure not to replace sugar with artificial sweeteners, which are just as bad or worse (see page 107). If you really need to sweeten something, consider chicory, which you can find at health food suppliers.

Simple Carbohydrates

This is really pretty much the same as stated above, as simple carbohydrates are sugars and cause the same problems. This category includes white flour (and anything made from it, such as bread and pasta), white rice, corn (which you should be avoiding because of its fungal content, anyway), and potato.

Even complex carbohydrates may have to be restricted (or temporarily eliminated, see Chapter 8, Cleanse), especially if you have more serious symptoms, since complex carbohydrates break down into simple ones—or complex sugars—and from there into straight glucose: sugar.

Refined and Processed Foods

You've got to skip the "junk food" (I hate to even dignify it by calling it food!). Yes, that means crisps and cookies and dough-nuts and just about anything you can get at fast food restaurants, and so on (and on, and on, and on). But it also includes many foods you may not have been concerned about before, such as the low-calorie frozen dinner you had last night, or the frozen burrito, or the canned soup. All these things are refined and processed to within an inch of their lives, and whatever nutrients they may have had to begin with are trashed in the process, even in the so-called "enriched" products. On top of that, they are loaded with sugar, salt, artificial colorings and flavorings, additives, preservatives, and butter, margarine, or hydrogenated or partially hydrogenated (hardened) vegetable oil—and deficient in fiber. They are, of course, acidifying.

Fruit

Though fruit has many good vitamins and minerals, and is rich in fiber, it is also filled with sugar. (Pineapples are 28 percent sugar, bananas 25 percent, honeydew melons 21 percent, mangoes 18 percent, apples 15 percent, oranges and cherries 12 percent, strawberries 11 percent, and watermelon 9 percent, just to name a few.) Despite what some nutritionists claim, there is no difference in your body between "natural" sugars and any other kind. Sugar is sugar—it doesn't matter if it is honey, jellybeans, maple syrup, or a berry or a piece of melon. In any form, microforms love it and will ferment it into alcohol and other mycotoxins and create an acidic environment in your body. So with the exception of lemons, limes, and occasionally nonsweet grapefruit, which actually turn out to be basic, fruit must be avoided to gain a healthfully balanced body. You can get all the same nutritional benefits from vegetables, without the negative side effects. Once you are in balance, a small portion of fresh, seasonal fruit, eaten by itself, can make a nice treat.

By the way, here's a little insight into why fruit gets sweeter as it ripens: The complex carbohydrates are fermenting into simpler and sweeter ones, which are then fermented further as yeast evolves and the fruit is actually turning into alcohol and mold—rotting, basically. Appetizing, huh?

Fruit juice is even worse, as the sugars are more concentrated (and the fiber is lost). Most of it is also processed and pasteurized, and almost always it is made from second-quality fruits—those that were too damaged or dirty or diseased to sell whole—and already contaminated with harmful microforms and mycotoxins.

THE pH OF FRUIT

The following is a list of common fruits with an approximate, relative potential of acidity (−) or alkalinity (+), as present in one ounce of food.

Rose hips	−15.5	Strawberry	−5.4
Pineapple	−12.6	Blueberry	−5.3
Mandarin orange	−11.5	Raspberry	−5.1
Banana, ripe	−10.1	Yellow plum	−4.9
Pear	−9.9	Italian plum	−4.9
Peach	−9.7	Date	−4.7
Apricot	−9.5	Cherry, sweet	−3.6
Papaya	−9.4	Cantaloupe melon	−2.5
Orange	−9.2	Red currant	−2.4
Mango	−8.7	Fig juice powder	−2.4
Tangerine	−8.5	Grapefruit	−1.7
Currant	−8.2	Watermelon	−1.0
Gooseberry, ripe	−7.7	Coconut, fresh	+0.5
Grape, ripe	−7.6	Cherry, sour	+3.5
Cranberry	−7.0	Banana, unripe	+4.8
Black currant	−6.1		

Dairy Products

Like most animal foods, dairy products contain hormone and pesticide residues, microforms, mycotoxins, and saturated fats. Layer on top of all those goodies milk sugar (lactose) that breaks down like any sugar and feeds harmful microforms. Dairy cows feed on stored grains laced with hormones and antibiotics made with fungi, which are then concentrated in milk. Then, too, cheese and yogurt are made by fermentation. And dairy is the leader of all foods in forming sticky mucus. It is highly acid-forming. It can increase cancer risk, including ovarian and endometrial. Furthermore, pasteurization destroys the beneficial enzymes milk starts out with. And pasteurization doesn't even really work! Pasteurized milk left out will rot and stink, whereas "raw" milk curdles naturally and is still edible.

With all that to recommend them, you can see why all dairy products should be eliminated from your diet. Try soy, almond, or rice milk as alternatives (though look carefully to avoid the vast majority of them that are filled with added sugar). If you must have milk, use unprocessed goat's milk from organically grown and grazing goats. It contains the antifungal caprylic acid.

No matter how many times you were told by teachers and parents to drink your milk, the idea that dairy products are healthy is pure hype—a cultural myth. Even if cows lived in some kind of bovine utopia and produced the perfect milk, let's face it: It simply isn't a human food. It is designed for baby cows, whose requirements are far different from those of humans. Milk is full of components of no use to us, and they must either be converted to use (wasting our body's resources in the process) or eliminated as toxins. No other animal species drinks milk beyond infancy—and certainly not from a species outside their own!

Milk is only the beginning of the problem. Consider that it takes 8 pints (4 liters) of milk to make one pound (450 grams) of hard cheese, 10 pints (5 liters) to make one pound of ice cream, and over 17 pints (8 liters) to make one pound of butter. Remembering that it takes 20 parts alkalinity to neutralize one part acidity, just imagine what it takes to counter the effects of so concentrated a source of acid! If it would take 20 cups of something alkaline to neutralize one cup of milk (already bad enough, don't you think?), you'd need 12 times as much—240 cups, or 12 gallons!—to neutralize a cup of ice cream.

No wonder so many people do so poorly on dairy foods. No wonder so many suffer with osteoporosis while still ingesting so much dairy. No wonder so many people have allergic reactions to dairy foods, or are lactose-intolerant. No wonder people can gain weight quickly on dairy foods and lose it so quickly when they go off these very concentrated foods. They are just too concentrated, and are ultra acidic in the bloodstream.

WHAT ABOUT CALCIUM?

We get asked this question a lot. It is true that calcium is vital for many functions in the body, but the current rage for getting huge doses of the mineral—through large quantities of dairy products daily as well as supplements—is based on faulty understandings of how the body uses it. Many people worry—*totally unnecessarily*—that if milk products are eliminated, their diet will leave them deficient in calcium.

The fact is that all leafy, green vegetables and grasses are inherently high in calcium (as well as iron, magnesium, vitamin C, and many of the B vitamins, but that's another story), as are celery, cauliflower, okra, onions, green beans, avocado, black beans, chickpeas, tofu, almonds, hazelnuts, and sesame seeds. In short, you get plenty of calcium with a diet that looks like the one described in this book. When we're asked about where we get our calcium, we often answer with a question of our own: Where does a cow get hers?

It is also important to evaluate how much calcium you really need to keep your bones and body healthy. To do so, you must understand that one of the things calcium does in the body is neutralize the acid created by eating animal protein. When you eat these acidic foods, the body tries to return to its alkaline state the only way it can—by withdrawing calcium from your bones if there isn't enough on hand in the food itself to do the job. Your kidneys also rob your bones in order to eliminate the excess nitrogen found in animal protein.

The current recommendations for 1,000 mg. a day of calcium and more assume an average American diet— which consists of one and a half to four times as much protein as necessary, creating an unnatural demand for calcium. Many experts blame the seeming epidemic of the bone-weakening disease osteoporosis on this protein overdose. It isn't really a lack of calcium at all! Or rather, it is a calcium-robbing problem, not a calcium-deficiency problem. We need to stop worrying about not getting enough calcium and pay attention instead to not getting too much protein. In the meantime, we're living the irony that getting plenty of calcium-rich dairy products can actually leave us with a negative calcium balance by the time all that protein is buffered.

Just to confirm, in a way that even mainstream science could understand, that we're getting enough calcium in our own bodies, we both recently had bone density tests. We have been essentially vegan for approximately twenty years now, and both of our tests came out with densities well above average. Shelley's rating was similar to a twenty-year-old's (when bone density usually peaks)—though she was forty-six at the time. Rob was also in the very highest percentile (and he was forty-eight).

CALCIUM CONTENT OF ALKALIZING FOODS
(PER 4 oz./100 g. OF FOOD)

Food	Calcium	Food	Calcium
Vegetables		Cabbage, red	42 mg.
Artichoke	51 mg.	Cauliflower	25 mg.
Asparagus	23 mg.	Celery	39 mg.
Aubergine	12 mg.	Chard, Swiss	88 mg.
Bamboo shoots	13 mg.	Chives	69 mg.
Beet greens	119 mg.	Collards (leaves)	250 mg.
Broccoli	103 mg.	Collards (stems)	203 mg.
Brussels sprouts	36 mg.	Cress	81 mg.
Cabbage, Chinese	43 mg.	Cucumber	25 mg.

Food	Calcium	Food	Calcium
Dandelion greens	187 mg.	Tomato, green	13 mg.
Fennel	100 mg.	Tomato, red	13 mg.
Garlic	29 mg.		
Kale (leaves)	249 mg.	**Legumes**	
Kale (stem)	179 mg.	Chickpea	150 mg.
Leek	52 mg.	Lentil, dried	79 mg.
Lettuce, Boston	35 mg.	Lima bean, fresh	52 mg.
Lettuce, iceberg	20 mg.	Mung sprouts	118 mg.
Lettuce, loose-leaf	68 mg.	Pea, green fresh	26 mg.
Mustard greens	183 mg.	Red bean, dried	110 mg.
Okra	92 mg.	Soybean, dried	226 mg.
Onion (green)	51 mg.	Soybean, fresh	67 mg.
Parsley	203 mg.	Soybean sprouts	48 mg.
Pepper, green	9 mg.		
Pepper, red	13 mg.	**Nuts and Seeds**	
Pepper, red hot	130 mg.	Almond	234 mg.
Radish	30 mg.	Brazil nut	186 mg.
Rhubarb	96 mg.	Hazelnut	209 mg.
Seaweed, agar	567 mg.	Pumpkin seed	51 mg.
Seaweed, dulse	296 mg.	Sesame seed	1,160 mg.
Spinach	93 mg.	Sunflower seed	120 mg.
Turnip greens	246 mg.		
Watercress	151 mg.	**Grains**	
		Barley	34 mg.
Fruits		Millet	20 mg.
Avocado (Haas.)	10 mg.	Rice, brown	32 mg.
Avocado (Ryan)	10 mg.	Wheat	46 mg.
Grapefruit, sour	16 mg.	Wheat bran	119 mg.
Lemon juice	7 mg.		

Salt

The negative effects of salt are well known, and yet the typical Western diet is crammed with it, starting with a salt-cellar probably on most dining tables. Even if you never use a salt-cellar, you can easily overdose on salt with processed foods—boxed, bottled, bagged, frozen, or canned—restaurant food, and junk food, all of which are, unless specifically labeled otherwise, generally loaded with salt.

Aim to eliminate all added processed salt from your diet, except alkalizing salts like Real Salt (an unprocessed natural salt mined in Utah, USA – see Resources) or seasoning salts with vegetables added.

Saturated and Other Unhealthy Fats

As healthful as the essential fatty acids are, the wrong fats are devastating. You know the litany of bad effects: Clogged arteries, heart disease, cancer, and so on. The villains are hydrogenated or partially hydrogenated (saturated, solidified) vegetable oil, margarine, butter, saturated fats, and almost all animal fats (from meat, poultry, eggs, and dairy—fish alone escapes).

Your aim here, too, is to eliminate all these dangerous fats from your diet. Don't cook your food in fat or oil (steam it), and don't smother it with them afterward.

I want to be clear, here, however: I am *not* advocating a non-fat diet. Your body needs good fats to survive, and to be fully healthy. Use good oils like olive, linseed (flax seed), and grape seed (and our favorites: Udo's and Essential Balance, which are blends!).

Meat and Eggs

Like dairy foods—like all animal products—meat (pork, beef, lamb, chicken, turkey, and so on) and eggs are filled with hormones, pesticides, steroids, antibiotics, microforms, mycotoxins, and the saturated fats that contribute to heart disease, strokes, and cancer, among many other things. (While I agree that the fat itself is part of the problem, consider also that fat is where animals' bodies store the toxins they've been exposed to.) And they are highly acidic. The animals feed on stored grain and pass along all the associated problems in their meat (see "Stored Grains" on page 101). What I'm trying to say is: Don't have anything to do with them.

There is a strong correlation between animal protein and several kinds of cancer, particularly breast, thyroid, prostate, pancreatic, endometrial, ovarian, stomach, and colon cancers. Studies show that people who get 70 percent of their protein from animal products have major health difficulties compared to those who get just 5 percent of their protein that way: 17 times the death rate from heart disease, for example, and five times the likelihood of dying of breast cancer (for women).

The consumption of eggs, alone, is associated with increased risk of colon cancer. I'm not surprised, as eggs from grain-fed chickens have been documented to contain mycotoxins. (My own observations have revealed that fifteen minutes after eating an egg, people will show bacteria, or an increase in bacteria, in their blood.) Dairy products were also incriminated in the same study, with the highest association being with cheese. Interestingly, increased consumption of red meat did *not* increase risk. I attribute this fact to the study being done in Argentina, where the beef cattle are usually pasture-grazed rather than grain-fed. This is not to say that red meat from grazing animals is good food—just that it may be the lesser of two evils.

An Australian study also turned up a positive association of egg consumption and colon cancer, as well as links with the intake of red meat, liver, dairy foods, and poultry.

Researchers studying the effects of a Western-style diet in Japanese women found that it was linked to a higher risk of breast cancer because of the much larger amounts of meat included. Scientists started down this track after noting that breast cancer was rare in Japanese women before World War II.

Another study linked poultry, ham, salami, bacon, and sausage to increased risk of thyroid cancer, as well as cheese, butter, and oils other than olive oil. (Olive oil is generally free of mycotoxins.)

Yet another study supported the fact that the type of dietary fat consumed influences the occurrence of endometrial, ovarian, and stomach cancers—with animal-derived fats contributing to an increased risk. People in the study who developed cancer ate more bacon and ham, used more butter in cooking, and drank more whole milk.

A Swedish study found a number of dietary factors to be associated with pancreatic cancer, including higher consumption of fried and grilled meat (as well as margarine on white bread—just to remind us that simply being vegetarian isn't enough to solve the whole problem).

Processed meats and cheeses are even worse, thanks to their nitrosamines, and are a risk factor for brain and spinal cord tumors.

Besides, animal foods are simply *dead*. Dead in every respect, including lack of enzymes. Vegetable foods, alive with enzymes, energy, and phytonutrients, are far superior in every way.

All meats properly aged for human consumption are, by definition, partially fermented, and thus permeated with microforms and their toxins. It is yeast, after all, that causes the aging, and the final taste and texture is determined by the nature of the microbial aging process. (That's on top of the mycotoxins in the animals' feed, which then show up in their muscles—the meat.) One of the specific mycotoxins involved has been linked to diabetes. In case you were wondering, most mycotoxins are heat-tolerant, so cooking doesn't get rid of them, even if it kills off some of their creators.

Whatever nutrients may be in animal foods, they simply are not worth the risks—not to mention the stress they put on the body during digestion and through the energy required to extract what nutrients they contain.

Anatomically and physiologically, humans are just not meant to be carnivores or omnivores. The long, complicated human digestive tract is designed for the slow absorption of complex and stable plant food. Carnivores have short, simple bowels to allow for minimum transit time of unstable, dead animal food. Their intestinal microorganisms are different from humans', too.

On the other side of the coin, starch digestion in humans is quite elaborate, whereas carnivores eat little or no starch. If humans were carnivores, we'd be sweating through our tongues instead of our skin. Flesh eaters have teeth and jaws designed for tearing apart freshly killed animals. Only our hand tools allow us to override this obvious natural limitation, not to mention the fact that we get none of the nutrition contained in fur, feathers, organs, and bones, the way true carnivores do. Finally, we seldom eat raw flesh. We almost always need to cook it to kill parasites and other harmful microforms, and to disguise the corpse that it is, none of which is necessary for real meat eaters. Humans are designed to be vegetarian, and our bodies will never work their best if we keep forcing them to do something they are not equipped to handle.

WHAT ABOUT PROTEIN?

We need to evaluate just how much protein we really need. Expert research suggests we need only 25 grams—just one ounce—a day of protein. The average Westerner who eats meat, eggs, and dairy probably gets 75 to 125 grams a day—three to five times more than we actually need. I believe protein should comprise roughly 5 to 7 percent of our total diet.

Our bodies are just 7 percent protein (and 70 percent water, 20 percent fat, 1 to 2 percent vitamins and minerals, and 0.5 to 1 percent sugar). Most meats are 20 to 25 percent protein—therefore providing more than the human body requires. If you don't eat meat, never fear: Spinach and other greens are higher in amino acids (the building blocks of protein) than steak!

Cow's milk, too, is protein-rich. In contrast, a protein source specifically designed for human consumption—breast milk—is only 5 percent protein (and some sources put it as low as 1.4 to 2.2 percent). And that's meant to be the sole source of nutrition for a human who is growing and developing faster than at any other time of life—doubling or tripling body mass and size within the first year of life. If we really needed super-concentrated proteins for good growth and health, surely mother's milk would contain a much higher percentage. As it is, I believe it reflects the body's actual requirements.

Some of the strongest animals in the world—take, for example, the gorilla, and the elephant—eat no meat. They are obviously not hurting for protein. What do they subsist on? Grass and leaves.

Stored Grains

Stored grains means last year's crop. Grains that are stored will usually begin to ferment within ninety days, and in short order are full of mycotoxins. They also harbor harmful microforms. So you want to get this year's crop, preferably within three months of being harvested. The only way I know of to get fresh grains is to ask the shop manager to check the dating on the packing (if you buy in bulk), or to check with the supplier. It is always important to read the labels.

Unstored grains are a healthy part of your diet, generally, though you should cut them out for the initial eight to twelve weeks at the beginning of our program—after a cleansing, or for as long as you have symptoms.

Eating *stored* grains is, not surprisingly, damaging to the body. For example, a 1991 study found positive correlations between eating stored grains and esophageal cancer. That same year, researchers identified cooked cereal (a form of stored grain) as a risk factor in stomach cancer.

Stored potatoes are similarly risky. To take just one example: In pregnant women who consume large amounts of potatoes, two mycotoxins produced by fungi commonly found in potatoes have been incriminated as a cause of spina bifida in their offspring.

Yeast

You must exclude all yeast—brewer's and baker's and "nutritional"—and yeast-containing foods from your diet. Obviously, you don't want to be taking in pure microforms. Besides, the most common ways to get yeasts are bad for you for other reasons, too: beer and wine (double whammies, as alcohol is a danger, too) and breads and baked goods (a triple whammy because of the stored grains the flour is made from and the sugar and other simple carbs they contain).

Eating yeast and anything made with yeast can spur microform overgrowth and increase mycotoxins (increasing the amount created in your own body, in addition to whatever's in the product itself). If you need anything else to discourage your use, you should know that yeast food products can cause kidney stones (and stones in the liver, gallbladder, and even brain), bone deposits, osteoarthritis, rheumatoid arthritis, kidney disease, heart disease, diabetes (in a 1990 study, all mice fed a diet containing 10 percent brewer's yeast developed diabetes), sarcoidosis (an autoimmune disease affecting the lungs, eyes, and skin), cirrhosis, and many cancers, particularly breast, prostate, and liver cancer. Other resulting symptoms include Crohn's disease and colitis.

Read labels carefully to make sure all your foods, condiments, and seasonings are yeast-free.

"Edible" Fungus

Mushrooms of all kinds and in all forms—besides the obvious problem that they are themselves the fruiting bodies of yeast or fungus—form acids as they are digested. They also contain mycotoxins that poison human cells and lead to degenerative diseases. In my opinion, there is no such thing as a good mushroom. The "edible" ones are just less poisonous than the ones that kill you immediately! Don't eat them, don't drink them, don't even sniff them. Mushrooms all contain various amounts of the mycotoxin amanitin, which, in large amounts, will kill you almost instantly. With smaller amounts the result is the same—it just takes a little longer.

In a 1979 study, a leading cancer researcher administered mushroom mycotoxins to mice in their drinking water. She noted *twenty-one different types* of cancer as a result. Now we know that all mushrooms contain at least five active ingredients that exhibit carcinogenic properties in animals.

Now many impressive health claims have been made for some mushrooms, but they do have occasional toxic side effects—and all the same problems as any other mushroom. I believe all of their alleged benefits can easily be obtained in other—safer—ways.

Spirulina and Algae Supplements

We wish we could go along with the glowing recommendations for spirulina and algae supplements. After all, they are green plants rich in chlorophyll, protein, minerals, and other nutrients. But they thrive in acid conditions. And just think of what they really are: the scum you see growing on the surface of stagnant ponds and lakes. Toxins in algae have been shown to harm the liver and nervous system, and one seems to spur tumor growth in animals.

Algae supplements do contain vitamin B12, which is not found in veggies—including this one in its pure state. Rather, it is made by bacteria that get into the algae via bird feathers and droppings. Call me crazy, but since we have our own intestinal bacteria that can make B12, I'd rather not get my B12 from bird droppings!

Fortunately, a day's dose of algae differs little from a serving of organic broccoli, so you can reap the benefits without facing the risks.

Fermented and Malted Products

This includes condiments such as vinegar, mustard, ketchup, steak sauce, soy sauce, tamari, mayonnaise, salad dressings, chili sauce, horseradish, miso, monosodium glutamate (MSG), and any kind of alcohol, as well as pickled vegetables such as relish, green olives, sauerkraut and, of course, pickles. And tempeh. All are acid-forming and create sticky mucus and, with the exception of MSG, are fermented by fungus.

Malt products such as malted milk and certain cereals and candy are also fermented by fungus and, besides containing high levels of sugar, are acid-forming and create sticky mucus.

Alcohol

It may help to think of alcohol as the mycotoxin made by yeast that it is. That includes wine, beer, whisky, brandy, gin, rum, and vodka, to name just the most popular. You already know abuse of alcohol causes disease, including cirrhosis of the liver, brain damage, cancers, fetal injury, and death. That's before you even factor in the damage any mycotoxin can do—and it doesn't take what mainstream medicine considers abusive quantities for serious harm to be done. On top of that, the liver can convert alcohol into yet another mycotoxin (acetaldehyde), with its own harmful ways.

Caffeine

The main sources of caffeine are chocolate, tea, fizzy drinks, and all forms of coffee—even "decaffeinated" contains enough to have a negative effect on you. All these things produce a lot of acid and a lot of mucus. Then think about the foods commonly eaten along with a cup of coffee or tea (often traditional breakfast foods and desserts), all of which are also acid-producers, and you'll see it is a recipe for disaster.

Furthermore, caffeine is addictive. You could take the word of researchers at Johns Hopkins School of Medicine for it—or simply observe your own headaches when you've been deprived of your morning jolt. Eighty-two percent of volunteers for that Johns Hopkins study showed withdrawal symptoms when they were given a placebo instead of their usual dose of caffeine. Official estimates are that more than 80 percent of adults in the United States regularly consume enough caffeine to produce addiction. Do your part to bring down that grim statistic!

Corn, Corn Products, Peanuts, and Peanut Products

Corn contains 25 different mycotoxin-producing fungi, including recognized carcinogens! Peanuts contain 26. On top of that, broken and ground nuts (of any kind) are ready targets for airborne mold spores and quickly become rancid. You can see it on the nuts as a dark or black discoloring. Contamination occurs during the growing process because the plants themselves are not resistant. Humans who eventually ingest them are also eating the fungi and their toxic waste, bombarding their digestive tracts with negative microforms.

Research has linked corn consumption with cancers of the esophagus and stomach, and peanuts with pancreatic and liver cancer.

Cashew nuts and dried coconut are similarly contaminated, and should also be avoided.

Heated Oils

Any oils that have been cooked, or heated in processing, have been nutritionally destroyed, including the biggest brands of corn, rapeseed, and other vegetable oils. Look for "cold-pressed" virgin oils instead, such as many olive oils, choosing from the healthy varieties, of course.

Microwaved Food

First of all, microwaving your food destroys enzymes, depleting the life energy, just because it is cooking it. But it gets much worse. The Russians, who have done the most diligent research on microwave ovens and their biological effects on food and humans, outlawed their use. In their research, foods that were exposed to microwave energy increased in cancer-causing effects and decreased in nutritional value. Vitamins and minerals were made useless in every food tested, and the bioavailability of nutrients, including the B vitamins, vitamins C and E, and essential minerals, decreased. Meat proteins were rendered worthless (not that we're recommending nonmicrowaved meat, but just to show you how powerful these machines are—in generally unacknowledged ways). Microwaving also interfered with the digestibility of fruits and vegetables. To top it all off, microwaving makes all foods acid-forming. In my own work, I see a higher-than-normal percentage of abnormal blood cells in the blood of people who eat microwaved foods.

Artificial Sweeteners

Artificial sweeteners are acidifying. They all break down into deadly acids in the body. For example, when you ingest aspartame, one of the ingredients, methyl alcohol, converts into formaldehyde, a deadly neurotoxin and known carcinogen! But that's not all. From there, it turns into formic acid (which is, by the way, the poison fire ants use in their attacks). And that's just one ingredient in one of the many artificial sweeteners.

A wide variety of symptoms can be caused by artificial sweeteners, including headaches, migraines, dizziness, vertigo, seizures, depression, fatigue, irritability, increased heart rate, heart palpitations, insomnia, vision problems, hearing loss, ringing in the ears, weight gain, numbness, muscle spasms, joint pain, breathing difficulties, anxiety attacks, slurred speech, and a loss of taste. Artificial sweeteners can also trigger or worsen arthritis, chronic fatigue, diabetes, fibromyalgia, brain tumors, MS, Parkinson's, Alzheimer's, systemic lupus, mental retardation, birth defects, thyroid disorders, lymphoma, and epilepsy. Don't let them into your body to do their worst.

Safer sweeteners to use would be natural plant sources such as chicory, which you can find in your natural food shop.

THE pH OF FOOD

The following is a list of common foods with an approximate, relative potential of acidity (-) or alkalinity (+), as present in one ounce of food.

Root Vegetables		Cream	−3.9
Corn	−9.6	Homogenized milk	−1.0
Stored potatoes	+2.0	Buttermilk	+1.3
Meat, Poultry, and Fish		**Bread, Biscuits**	
Pork	−38.0	**(Stored Grains/**	
Veal	−35.0	**Risen Dough)**	
Beef	−34.5	White bread	−10.0
Ocean fish	−20.0	White biscuit	−6.5
Chicken	−18.0 to −22.0	Wholemeal bread	−6.5
Eggs	−18.0 to −22.0	Wholegrain bread	−4.5
Oysters	−5.0	Rye bread	−2.5
Liver	−3.0		
Organ meats	−3.0	**Nuts**	
		Pistachios	−16.6
Milk and Milk Products		Peanuts	−12.8
Hard cheese	−18.1	Macadamia	−11.7
Quark	−17.3	Cashews	−9.3

Fats

Margarine	−7.6
Corn oil	−6.5
Butter	−3.9

Sweets

Artificial sweeteners	−26.5
White sugar (refined cane sugar)	−17.6
Beet sugar	−15.1
Molasses	−14.6
Fructose	−9.5
Partially refined sugar	−9.5
Milk sugar	−9.4
Barley malt syrup	−9.3

Brown rice syrup	−8.7
Honey	−7.6

Condiments

Vinegar	−39.4
Soy sauce	−36.2
Mustard	−19.2
Mayonnaise	−12.5
Ketchup	−12.4

Beverages

Spirits	−28.6 to −38.7
Fruit juice sweetened with white sugar	−33.4
Tea (black)	−27.1
Beer	−26.8
Coffee	−25.1
Wine	−16.4
Fruit juice, packaged, natural	−8.7

Chapter 5

You Are What You Drink: Water, Juice, and "Green Drink"

At least as important as how you eat is how you *drink,* beginning with the fact that most of us simply don't drink enough. Then, when we do drink, most often we don't drink what is good for us. At the very core of this program are three simple strategies to change all that.

WATER

Here's the thing: Water is of the utmost importance to becoming and remaining healthy. But water quality, already atrocious, will be inexorably deteriorating for the foreseeable future.

Like the earth on which we live, our bodies are 70 percent water (and our blood 94 percent). If we subsist on polluted water, imagine the devastation to our bodies. Come to think of it, you don't have to "imagine"—chances are you are experiencing it right now.

The single most important thing you will learn in this book is to get your body plenty of pure water. And not just any water—alkalizing water. Ideally, you'll get at least four liters (about one gallon) of good water every day. If that sounds like a lot to you right now, don't worry: As you hydrate your body properly, you'll develop more of a thirst for water. You should also note that food cravings are often the body's cry for water. You might already have more of a thirst than you even realize.

Getting liberal amounts of alkaline water (having a pH between 9 and 11) neutralizes stored acid wastes and, if consumed every day in conjunction with a good diet, gently removes the acids from the body.

I'm sure you won't be surprised to learn that the water running from your tap—even if you filter it—is not healthy. Most municipal water supplies in the USA are a disgrace, especially those poisoned with chlorine and fluoride, which means most of them (In the UK, you can request a chemical analysis of your water supply from your water company or OFFWAT—see local telephone directory). Bottled water, though usually better-tasting, may also contain many impurities, or simply be dead from processing and storage. In the USA, commercial drinking water standards ignore thousands of potential pollutants. The Environmental Protection Agency lists around 200 primary, major water pollutants for which municipal and commercial drinking water must be tested. There are thousands more unidentified, and thousands more that are variants or combinations (but not listed). No one can screen for all possible poisons in all water supplies. Some testing procedures in the USA are inadequate, and some are very expensive: Tests for some of the worst contaminants run up to 1200 dollars—*for each separate chemical!* And do not be fooled by taste. Some of the deadliest pollutants are tasteless (one reason official standards can ignore them).

But don't despair. Plain tap water is out, but you can get good water from your own sink through distilling or reverse-osmosis (purified) filtering. Distillation is the evaporation and condensation of water. Distilled water comes closest to rainwater, which, if our atmosphere wasn't so polluted, would be the ideal source of water. Reverse-osmosis is a multifilter process that purifies water of toxic chemicals and large mineral deposits. Distilled and reverse-osmosis water—like rainwater—have more oxygen atoms or hydroxyl ions (OH^-), and fewer hydrogen ions (H^+). More hydrogen makes water acidic, so this way the water is neutralized and can help our bodies reduce accumulated acid wastes. Scientific explanations aside, with the right equipment installed under your sink, healthy water ready to drink comes right out of your tap.

Both processes make the water more neutral, and it can then be made more alkaline as necessary with the addition of pH drops, such as hydrogen peroxide (H_2O_2)—five drops per eight to 12 fluid ounces/350 ml. of purified water. Look for regular hydrogen peroxide or Aerobic Oxygen at the chemist.

When added to the pure, neutral water you drink, and thus added to your bloodstream, pH drops act as an oxygen catalyst, alkalizing, neutralizing, oxygenating, and pH balancing the body. Allow an hour or two after a heavy protein meal before drinking water with pH drops, so you don't interfere with the stomach acid digesting the protein.

I only know of one home appliance in the USA that provides a reasonable simulation of nature's water-processing system, alternately heating and cooling the water. It is called the Living Water Machine (see Resources). It produces very pure distilled water that is said to be biologically active—that is, alive—by keeping out some organic solvents that other machines may transfer to, and even concentrate in, the distilled water. It also stirs the water with a small fan to oxygenate it further. As a bonus, it is relatively easy to clean. It is, I must warn you, an expensive unit. Fortunately, reverse-osmosis systems, which attach right under your sink, can be had at a reasonable price, and are a good solution if you can't afford distillation.

Whichever method you choose—buying distilled water, distilling it yourself, reverse-osmosis filters, drops—just stock your fridge or pantry with good water. And drink up! Drinking between meals is especially important. On this program you may not feel a need to drink with meals, because so many of the vegetables you'll be getting contain so much water—many of them are 70 to 90 percent water.

We like to squeeze fresh lemon or lime juice in our drinking water to boost its alkalizing effects (and it is tasty, too, of course).

JUICE

Eight fluid ounces of fresh vegetable juice is an ideal beginning to any meal or an excellent snack. All the benefits of vegetables (and grasses) can be enhanced simply by juicing them. The nutrients are more concentrated and more quickly and easily available to the body. You do lose the fiber with juicing, but that is what frees the nutrients. (Chewing does the same thing, just not as completely as juicing does.)

You do need fiber, so you wouldn't want to get *all* your vegetables this way. But when you "drink your vegetables," your body is getting a greater concentration of rapidly usable alkaline salts, vitamins, minerals, chlorophyll, and enzymes, so vegetable juices are very alkalizing. They also have an important cleansing effect in the intestines. Juicing vegetables you might otherwise cook also provides enzymatic relief to the digestive organs.

The best, most alkalizing juice is made up mostly of green vegetables and grasses. (Fruit juices must be avoided, especially early in this program, because of their large amounts of sugars.) At first, when your tastebuds may not yet be accustomed to the more subtle sweetness found in greens, you may want to add some carrot, beet, or red, yellow, or orange bell peppers. You can even use butternut squash and sweet potato. These vegetables are sweet because they have higher levels of sugar, so use them moderately, keeping them to 20 percent or less of your juice—meaning 80 percent green. (Beet can also be a vigorous lower bowel cleanser, another reason to go easy on it, especially at first.) As your body gets more basic, green juices will taste better and better to you, and then you may want to lower the proportion of carrot or beet juice to 10 percent or less to get even more greens. (Peppers are not high in sugars, so you can use them freely.)

The recipe section provides many ideas for juice combinations, and the Resources section includes some good juicing books if you want to explore more. In addition, all juicers come with some recipes. Experiment a bit to find the combinations you enjoy most. All green vegetables are terrific for juicing (we use a lot of celery, cucumber, broccoli, green bell pepper, courgette, green beans, lettuce, cabbage, beet greens, and leafy greens of all kinds). Tomatoes are also great for juicing—as you probably already know. But don't rely on processed and canned stuff—make your own! Juicing grasses is a good way to reap their benefits, without feeling as if you are chewing your cud. Juice sprouts for a kind of double bonus, concentrating still further the nutrients that are already so dense and making them even more alkalizing than they are whole. Also think about spicing up your juice with arrowroot (very mild), parsley, radish, ginger, and garlic.

Because of the way juice concentrates everything, using quality produce is particularly important here. Use organic whenever you possibly can, and, as always, get it and use it as fresh as you possibly can. Wash it well, particularly anything that's not organic, and perhaps soak it in pure alkaline water (twenty drops to the gallon of Aerobic Oxygen or hydrogen peroxide).

Peel heavily waxed vegetables. But there are plentiful nutrients in the skin, so leave them on whenever you can. Don't forget to use green vegetable tops, too, such as beet and carrot greens. When you can't get organic or newly picked vegetables, you can bolster your juice by adding some dried, powdered "greens" (like the ones described in Chapter 10, Supplements). You can use wheat grass juice powder as an alternative to juicing wheat grass itself (if it is difficult for you to buy fresh, or if your juicer isn't up to the task).

When you spin, whip, shake, and press vegetables into juice, the microzymas excrete acid waste, making the juice mildly acidic despite coming from alkaline vegetables. So make your fresh juices highly alkaline by diluting them with distilled water (just one part juice to ten to twenty parts water), then adding pH drops (10–20 drops for 10–20 fluid ounces/300–600 ml.).

Making your own juice is the best option, for freshness. And drink it up as soon as you make it. Don't let it sit for more than a few minutes. (If it is going to sit more than ten or fifteen minutes, maintain its goodness by adding three to five drops of colloidal vitamin C to the juice container as it begins to fill, or dissolve 250–500 mg. crystalline vitamin C in a few ounces of pure water and add that to the juice container before beginning.)

Your own juice will always be better than packaged, preserved products. Pasteurizing juice—and almost all of it, even in health food shops, is pasteurized—evaporates the enzymes and destroys the life force.

Take the time and trouble to choose a quality juicer—it's worth it. You'll find a wide range of prices, though more expensive does not necessarily mean better. If possible, talk to owners of different machines and see what they think—and see if they'll let you test drive theirs. Look for an efficient one that has the ability to juice continuously—stay away from hand-press juicers and hand wheat grass juicers—and is easy to clean (has few moving parts and disassembles and reassembles easily). To juice grasses or parsley, you'll need a high-powered juicer (check the packaging for voltage and wattage). There's even one juicer available (the Green Power® Juice Extractor), which ionizes the juice, which is a bonus in my opinion. It also does a beautiful job with grasses, which not every juicer can claim (see Resources). That machine does not heat the juice, as some do. Make sure you get a machine that does not heat the juice! A juicer that whips or spins the juice will cause it to heat up as the molecules bounce off one another. As with pasteurization, heat causes the enzymes to evaporate and reduces or destroys the life force. My juicer is rotary-geared, and has a gravity fall for the juice (versus a centrifugal spinning juicer that needs the filter changed).

JUAN'S STORY

Two years ago, I came down with what I thought was the flu—until I noticed blood in my urine. I got myself checked out by my doctor, who told me my kidneys were failing and admitted me to the hospital. I went home a week later with the diagnosis of kidney disease, though I was told I would pretty much be able to go back to my regular life. But as the months went by, I just got sicker and sicker. Going to work and doing my normal daily activities felt almost impossible. I was in and out of the hospital four times, and more than once I didn't think I'd make it out alive. Even on my good days, I felt like I was slowly dying.

Two days after I first heard about The pH Miracle program (and before I'd done anything about what I'd heard), I collapsed in church with a fever of 104 degrees and was rushed to the hospital. My wife ordered concentrated green powder and pH drops—although my doctors dismissed them out of hand—which I started on as soon as I got home. After about a week, I started to have a little more strength, and I made a commitment to changing my life and undertook the whole program. With a radically new diet, I saw incredible results within a month. I had no more blood in my urine; I wasn't tired all the time; I felt up to playing with my children. It's been almost a year and a half since the last time I was in the hospital. I have never felt better! At a checkup two weeks ago, my doctors couldn't find a thing wrong with me or my kidneys. They don't know how to explain it, but I do!

I want to note quickly another food-processing option that at a glance seems similar to juicing: The purée method, where a special machine whips whole fruits, vegetables, and even grains into liquids or creamy forms. But this process mixes a lot of air with the food, which is not welcome in the stomach. It does leave the fiber in, and fiber is good for you. On the other hand, it leaves the fiber in—which makes the nutrients less accessible to your body. The creamy consistency also discourages chewing and encourages faster eating, thereby depriving food of the oral secretions necessary for full and proper digestion. Together with the large amounts of solid matter, this places greater stress on the digestive system than does juice.

So, while there may be some benefits to this method, don't substitute it for juicing. Avoid it altogether as you start on this program, and if you decide to add it later, don't use it at the same meal as juice.

"GREEN DRINK"

Here's a way to take what you get out of proper hydration, and what you get out of vegetables and vegetable juices, and go one better: Green powders. These are grasses, sprouted grains, and green vegetables, dried and powdered and sold as supplements. They infuse your body with easily absorbed vitamins, minerals, and amino acids (the building blocks of proteins). Make sure you get a product that is organically grown.

As mentioned, you can add some to your vegetable juice for extra "oomph." We usually just add green powders to pure water (often with pH drops) as part of our daily hydration—"green drink" made by blending one teaspoon into one liter (about two pints) of water—that is, three teaspoons into three liters of water for the day, plus, since you want to get at least four liters of water a day, some plain water in addition to the green drink. Just make sure you avoid green powders with algae, mushrooms, or probiotics. The probiotics are bacteria and sometimes ferment the green grasses in the formula, making it acidic. Probiotics are helpful supplements, but only when used properly on their own (see Chapter 10).

Just keep your green drink in a water bottle so you can shake it up to keep the greens well mixed. Make up one bottle at a time, and use at room temperature. Your body has to work to warm up an iced drink or cool down a hot drink, so you can save it some stress this way.

DAILY HYDRATION SCHEDULE

- Upon rising: 2 pints/1 liter of water with hydrogen peroxide (with lemon or lime juice as desired).
- Between breakfast and lunch: 3 pints/1½ liters of water with 1½ teaspoons of green powder and 24 drops of Aerobic Oxygen/hydrogen peroxide.
- Between lunch and dinner: 3 pints/1½ liters of water with 1½ teaspoons of green powder and 24 drops of Aerobic Oxygen/hydrogen peroxide.
- Between dinner and bed: water as desired, with lemon/lime juice, and eight drops of Aerobic Oxygen/hydrogen peroxide per 2 pints/1liter.

Many people ask me about what good alkaline water does since "everyone knows" it is just going into the highly acidic stomach. The problem here is not with water meeting acid, it is with what "everyone knows." There is no hydrochloric acid pouch in our body. The stomach wall makes HCl instantly, on an as-needed basis. "As needed" is determined primarily by what the stomach is sent. Low water content, acid-forming foods like meats, eggs, and breads cause the release of larger amounts of HCl, in order to break them down. High-water-content foods such as nonstarchy vegetables require much less HCl. Water—being extremely high in water content!—does not trigger the release of HCl, so it does not encounter an acidic stomach environment if taken on its own or with alkaline meals. So drink up!

Chapter 6

Food Combining

To ensure thorough and proper digestion, food combining is an important consideration. And there's a lot out there designed to help you understand and implement various food-combining systems. While the idea is key, however, the vast majority of available programs are usually confusing, are often inaccurate, and tend to offer conflicting advice. And they are all too unnecessarily complicated.

I'm here to tell you it doesn't have to be that way. The thing to remember is that the human digestive system is not designed for complex meals. Different foods make different, specific demands on the digestive system. That we are capable of digesting many different kinds of foods doesn't mean we can do so all at once. For example, protein digestion requires a highly acid environment and takes place in the stomach. In stark contrast, starch requires a mildly alkaline environment for digestion, which takes place in the mouth and small intestine. The same is true for vegetables. (Fats also require a mild alkaline environment and are digested in the small intestine.)

It doesn't take much to imagine that protein and starchy foods do not do well when eaten at the same time. One will interfere with digestion of the other, causing incomplete digestion of both. Whatever is not efficiently digested by you will be "digested" by harmful microforms. It's another vicious circle: Compromised digestion paves the way for negative microforms, and negative microforms further disrupt digestion. Poor food combining is also a major cause of the formation of sticky mucus.

Take a minute to stop and think of all the American "classics" that combine protein and starch—meat and potatoes, fish and chips, chicken and rice, a burger and fries, ham sandwich (any kind of sandwich), to name just a few—and you'll begin to realize just how badly we've abused our digestive systems. Most of us don't even know what it would be like to have proper digestion!

LUCY'S STORY

Because of my family history of sky-high cholesterol levels, and the terrible heart consequences, I'd always been careful about what I ate. As a home economist, homemaker, and mother of eight children, I was also careful about what I served my family. As my health deteriorated, with all kinds of symptoms bothering me, I experimented with different "healthy" ways of eating, constantly fine-tuning my approach.

I grew up eating well. My mother followed the Food and Drug Administration's (FDA) recommendations of the time, serving vegetables, whole grains, lean meats, and fresh fruits daily. Through my own early years as a mother, I moved to less meat, chose fresh vegetables—and usually served them steamed—switched to brown rice, and started to use natural supplements. I ground whole wheat myself and made fresh bread weekly. I eliminated fizzy drinks, simple sugar, processed foods, and milk. My health improved some, but not completely.

My health really began spiraling down after the birth of my eighth child, which required an emergency C-section and two blood transfusions far from home. As more negative symptoms appeared, and I felt my energy and vitality slowly ebbing, I worked harder to unlock what good nutrition could offer me. Food combining was one of the first things I explored, but my early results were discouraging.

I've tried different strategies over the years. I began by using the four basic food groups. A typical dinner was baked chicken, pan-fried potatoes, frozen broccoli, a canned peach with cottage cheese, and oatmeal cake. I experienced a full feeling afterward and felt as if I would like to lie down for a nice nap. And I continued to have hypoglycemia, high cholesterol, and sinus infections, among many other things. Next, I added more whole grains and fresh vegetables and cut back on meat, as suggested in the current FDA "food pyramid." A typical dinner was brown rice and chicken casserole, fresh steamed broccoli, a slice of homemade whole wheat bread with butter, and homemade applesauce. My blood sugar stabilized, but I continued to have cravings as well as a wide variety of other health issues.

Next, I tried eating nothing but fruits and fruit juices from dawn until noon. The rest of the day I would be careful to have only one "concentrated" food (protein or fat) at a meal—and no more fruit. I ate meat, but never with a starch. The large amounts of fruit kept me craving sweets, and I experienced low periods every afternoon. I never felt energetic after eating meat. And I didn't find the food satisfying. So I went back to my previous diet—and gained more weight and added a host of health concerns.

And so it went on until I learned about the Youngs' program, and the proper way to combine foods. The day I started eating alkaline and drinking a gallon of water with pH drops and concentrated green powder every day, my life changed. I immediately noticed a rise in my energy levels. The most important changes I made, besides getting plenty of good water, were to eat something raw at each meal, focus mainly on green vegetables, and use the more alkaline grains.

At first I wanted that feeling of having something stick to my ribs, but found that the high-water/low-sugar foods gave a sustained energy that I was not used to. Now dinner is typically a vegetable, soba noodle, and tofu stir fry, or occasionally, a small portion of grilled salmon, Jasmine rice with almonds, fresh steamed asparagus, and some raw pepper strips. For lunch I almost always have a fresh salad built from spinach, dark green lettuces, avocado, cucumber, celery, carrots, radishes, pumpkin seeds, sprouts, a little baked tofu, and a dressing of lemon juice, olive oil, and spices. (For years I had cut out all fat/oil of any kind due to my cholesterol challenge. Nothing helped until I added liberal amounts of the good essential oils to my diet daily.) To that I add a vegetable and hummus wrap in a sprouted wheat tortilla, or a brown rice cake with almond butter. That follows a breakfast of steamed millet with avocado, tomatoes, and flax oil, or lightly steamed broccoli and buckwheat cereal. That usually keeps me going strong until well into the afternoon. I never feel low in the afternoon anymore.

Sometimes I snack on a handful of soaked almonds. Often I make a gently warmed vegetable soup with an organic vegetable broth, which I enjoy for breakfast, lunch, or dinner. These foods give me all the energy my body needs. The foods I crave now are healthy, alkaline foods that are high in water content and low in sugar.

I now understand that though we all have our genetic tendencies, we are not bound by them. The gene might be the bullet, but the trigger is our lifestyle. This lifestyle has proven successful for me and my family for over two years now, keeping us healthy, energetic, and satisfied. We find the food delicious and sustaining. I have enjoyed developing recipes that are as healthy and beautiful as they are tasty. I prepare alkaline meals for my family of five every day, and on Sundays, when the rest of our family and friends come to eat, I serve alkaline food for twenty or more. They've enjoyed it so much that my married children have adopted some of these principles for their own young families, and they've all enjoyed health benefits. I feel my quest for a truly healthy way of eating has finally paid off.

Combining sugar and starch, or sugar and protein, leads to the same kinds of problems. Fortunately, the solution is simple: Mix no more than four foods, from no more than two types of food, at any given meal. For example, have steamed broccoli, and a green leaf (e.g. mesclun) and tomato salad with marinated tofu *or* soba noodles, but not both (three vegetables, and one protein, or three vegetables and one complex carb). Choosing fewer foods provides the simplest load on the digestive system. With that in mind, and following the general principles of this program, if you use only one protein per meal, and only one complex carbohydrate per meal, you are most of the way there.

Pay careful attention to food combining in the initial weeks after a Cleanse (see Chapter 8). Once you're fully on an alkalizing diet, it gets simpler still. When you're eating mainly foods that are high in water and low in sugar, you no longer need to worry about proper combining. You can't help but combine them properly, since you are, for the most part, limiting or eliminating the problematic foods. The foods that have makeups most similar to that of our bodies (high water content, 70 percent or more; naturally occurring oils, 20 to 30 percent; low protein, 5 to 7 percent; and even lower sugar, 0.5 to 3 percent) all combine with each other with no problem.

When you are strong and symptom-free, you can indulge in more complex meals with no real harm. Still, at the beginning of the program, or if you are seriously ill, or if you just want to ensure you're on an ideal regimen, paying strict attention to the rules that follow will serve you well. (See Resources for some books and charts with more details about food combining.)

FOOD-COMBINING BASICS

All you're trying to do is keep starch and animal protein separate, and keep sugars, including fruit, away from just about everything. That's why, once you have fully made the transition to this program, avoiding animal proteins, sugars, and most fruits, combining is not an issue. In the meantime, here are the official rules:

1. Low-sugar/high-water vegetables (or fruits) combine with everything. Eat them with protein, starch, or cold-pressed oils—and with other vegetables!

2. Eat starches with vegetables or low-sugar fruits. Don't eat starches (including starchy vegetables) with animal protein, acids, fruit, or oil. (For the purposes of food combining, "acid" is not necessarily the same as foods that make the body more acidic. The two most important examples of this exception are lemons and tomatoes, which are themselves acidic but which actually make the body more basic.) So, when you do choose a grain (including bread or pasta) or winter squash or potato, eat it alongside vegetables, and not with fish, for example.

3. Eat animal protein with vegetables or low-sugar fruits. Don't eat animal protein with starch, acids, or oils. Vegetable proteins combine with all low-sugar, high-water-content vegetables and fruits, as well as with good oils. Here's the flip side of the point above: When you're having fish, serve it with vegetables but not a grain. Get over paella (fish with rice); try fish on a bed of steamed greens, or atop a crunchy salad.

NOTABLE EXCEPTIONS:
AVOCADO AND TOMATO

Avocado is actually a fruit, but because it is low in sugar and relatively high in protein, it can be combined with vegetables, even the starchy ones, as well as with grains. So, I enjoy an avocado sandwich on yeast-free spelt bread, or avocado and tomato slices with lemon juice on jasmine rice.

Tomato too is a fruit. And although it is acidic, it has an alkaline effect in the body because of its low sugar content. So, like the avocado, it can be combined as if it were a vegetable.

4. Eat high-sugar fruit on its own—if you eat it at all. (Don't eat fruit with protein, starch, vegetables, or oil. In fact, don't use fruit at all—with the exception of lemon, lime, raw tomato, avocado, red, yellow, green, and orange peppers, and nonsweet grapefruit—unless you are quite well, and then only in moderation and in season.)

5. Eat (healthy) oils with vegetables and low-sugar fruits (tomato, avocado, red, yellow, orange, and green bell peppers, lemon, and lime). They also combine with starches (which must be kept to 20 percent or less of your diet). Do not eat healthy oils with animal fats or proteins. Seeds, nuts, and avocado—all excellent sources of healthy fats—can be combined with plant proteins, starches, or even high-sugar fruits. Don't douse your fish with oil or butter—use lemon juice, salsa, or herbs instead—and you'll be all set.

Now, for the "whys" of those wherefores:

1. Most vegetables, and the few fruits we've mentioned, are your healthiest choices anyway, and the fact that they combine with any other healthy choice just makes them even more ideal as the focus of your diet.

2. Starch and animal proteins are a bad combination, as explained earlier. Acids block the action of ptyalin, a component of saliva that is necessary for proper starch digestion. Starches, such as potatoes, bread, or pasta (and even whole grains), break down into simple sugars in the body, so adding high-sugar fruits just layers sugar on top of sugar—and acid on top of acid. The combination creates enough poisons that it can actually shut down the immune system for five hours—or even longer. Oil slows digestion of starch—though this won't be a problem if the starch is no more than 20 percent of your otherwise alkalizing meal. Oil can neutralize acids, so you don't want to have to avoid the healthy ones.

3. When animal protein is digested in the stomach, it creates acid. When combined with starches, the sugars in the starches make even more acid, leading to indigestion, heartburn, and wind—on top of all the other negative effects of a body that's too acidic. The same thing happens when you add more acid (including the acids created from the digestion of high-sugar fruit). Oils slow the digestion of animal protein, causing constipation and eventually acid reflux, heartburn, and wind.

4. Fruit—at least, the vast majority of fruit—is high in sugar, and acid-forming, so it is problematic even on its own. Combined with protein it is a recipe for excess

acid (as well as indigestion and wind). Starch and fruit is just double the sugar. In addition they have vastly different digestion times (fruit is digested extremely rapidly), opening the door to fermentation right in your digestive tract. Mixing fruit with oil can lead to constipation and poor absorption of nutrients. Finally, while fruits are cleansers, vegetables are builders. You unduly stress your body by asking it to do opposites simultaneously.

5. Oil slows the digestion of animal proteins and starch (though the latter will only be a problem if your starches are exceeding that 20 percent of your diet).

A FEW NOTES

Lemons and limes, or lemon and lime juice, are commonly thought of as acidic, but they actually have an alkaline effect in the body. So they do not fall under the warnings against combining with acids, and can be used together with starches, proteins, or oils.

SEPARATING FOOD AND DRINK

One other combination that can be bad in terms of digestion is food and beverages, even water. Don't wash down food with a drink. Cold drinks are particularly troublesome, as cold shuts down digestive activity as easily as it preserves food. Water (or other liquid) dilutes digestive chemicals, so it should be drunk at least half an hour before, or one hour after, a meal that includes animal protein. If you are eating a strictly vegetarian meal, feel free to drink along with it. We recommend eating juicier food items first, such as vegetables and salads, to pave the way for heavier items later in the meal. You may also find that a few sips of warm water after a meal aid digestion.

Part II

THE
PROGRAM

The chapters in this section provide all the details on how to embark on The pH Miracle program and how to incorporate it into your life, whether you choose to go all out immediately to address an urgent situation, or take a more deliberate pace for permanent change. So right here I'm just going to give you the four major steps of the program, without going into a lot of detail. I want you to see the forest before you get too involved with the trees.

The most basic outline of the program looks like this:

Step 1: Transition, for twelve weeks, gradually substituting for and eliminating acidifying foods, switching over to an alkaline diet.

Step 2: Cleanse, for one week, with some supplements and mild, natural laxatives.

Step 3: Strictly alkaline, for seven weeks, with additional supplements, including probiotics. This means eating only alkaline foods. Whereas your ultimate diet will be 70 or 80 percent alkalizing, at the outset you're aiming for 100 percent. Essentially, you stick to high-water-content vegetables, mainly green ones, and have them raw as much as possible.

Step 4: Maintenance. Now you can proceed on to 70 or 80 percent of your diet being alkaline and adding the full range of healthy foods, including fish, grains, soy, and starchy vegetables. You'll also add your full range of supplements, including any you need to address specific symptoms.

For the purposes of writing a book, we've settled on general guidelines everyone can follow. But truly the program should be individualized. Listen to your body and start where it is telling you you need to be. Observe how it changes as you go through the steps and pace yourself accordingly. There's nothing magic about the timing of these four steps, and extra weeks here or faster progress there is all part of the experience, as long as you stay basically on track. The pH Miracle works a bit differently for everyone who follows it, though the end result—radiant good health—is the same for all.

If you are very ill, you might start directly with step 2 to get the fastest results, and follow the strictest part of the program (step 3) for up to three months. Those of you who begin not *too* far out of balance may need only one month in step 3 before going with the fullest range of foods. It all depends on your progress.

You can also begin with a Cleanse before you begin transitioning, if you want to give yourself a kind of jump start, but be sure to come back to the Cleanse again as you are finishing your transition and ready to go ahead with the strictly alkaline diet. Some or all of the steps of the transition may take you longer than a week—go ahead and do what you need to do. It is a lifetime investment you are making, so doing it right is far more important than doing it quickly.

I've seen this program work, time and again, for seriously ill people—as well as for people who seemed pretty healthy in the first place. What works for one person, or even most people, however, may not do so for everyone. You must tune into your body, its needs and responses, and take personal responsibility for it. Ultimately, no one knows more about you than you. This program is designed for people to manage on their own, but it is wise to seek input from knowledgeable practitioners if it seems appropriate. (If you find you are having trouble communicating your situation fully, you might ask the professionals you consult to have a look at this book.) In any event, you should consult a health care practitioner before beginning this (or any other) diet program.

Chapter 7

Transitioning

To be able to assimilate your food more readily, you must take small bites. Quite literally, you should not bite off more than you can chew. So it is with making the transition to an alkaline diet. Ease into it if necessary, with a series of small victories, rather than trying to master the whole thing at once. Don't discourage yourself by trying to change too much too quickly. Make changes gradually. That is generally best for the body anyway, and increases the chances that you'll succeed in the first place *and* the chances that you'll stick with it for the long haul. (The exception would be serious illness, when a drastic change may be just what you need—or when there may be no time for a stepwise approach.) Moving toward an alkaline lifestyle is a process—not a single event or an overnight transformation. As you make your way "home," enjoy your journey.

Our family took more than two years to make the transition complete (and my 13 year old is still transitioning). In part, that's because we were working out the system as we went along. On the other hand, we were already almost always vegetarian, so the total change was not as dramatic as it might have been starting from a more typical Western diet. Whatever the specific time frame that works for you, take it step by step, as we did, for clear—and lasting—results. Work through one transition at a time, allowing at least a week, and up to two to three weeks if you need to, to get acclimatized at each step. Or take on a few together if that feels comfortable. Move on when you feel at home with them. Feel free to change the order of the transitions. You'll be building a solid foundation, and then with layer upon sturdy layer on top—it'll be built to last.

Here's our very own 12 step program:

TRANSITION 1: BREAKFAST

Probably the single biggest change you will make on this program is in what you have for breakfast. So it is as good a place to start as any!

Most Westerners need to have a change of heart and mind concerning breakfast. Almost all of the conventional choices—eggs, pancakes, syrup, hot or cold cereals, fruit, juice, coffee, yogurt, bread products, sausage, bacon—make your body acidic or promote (or contain!) yeast or fungus or other microforms. Many contain huge amounts of sugars and simple carbohydrates, which acidify the blood and tissues, creating the environment that promotes the microforms. Others are dense sources of protein (and, almost always, fat), which, in addition to being high in parasite activity, also promote microform overgrowth. And all these acidic foods are also very low in water content—and extremely constipating. It's no wonder laxatives are one of the best-selling over-the-counter remedies. On top of all that, we eat them in dreadful combinations (eggs and fried bread, cereal and milk, toast and jam). What a way to start the day! Your body deserves to be replenished much more gently and wholesomely after the night fast.

So don't let the first meal of your day slow you down. (And I mean really slow you down—these acidic breakfast foods are also very low in water and constipating.) This basically means making the same choices at breakfast that you would at any other time of the day. It may seem strange at first, but you'll be doing yourself a big favor by switching over to soup, say, or a veggie wrap, or salad. Or how about a big plate of steamed broccoli? Or a colorful veggie juice? My (Shelley's) favorite is the Zippy Breakfast (see page 328) made with buckwheat (a seed) instead of a starchy grain. We need to learn from the traditions of some other cultures—for example, you'd be offered soup for breakfast in Japan. When we traveled in Israel, we were delighted to see tomato-cucumber salads as a part of every breakfast table. The American and European way—tremendous doses of sugar and protein, not to mention a big dollop of caffeine—might give you a short shot of energy initially, but over the long term the negative impact is drastic.

So begin with this new breakfast strategy, starting your day with a low-carbohydrate, high-fiber, high-water-content—and delicious—meal. Try it even just for a couple weeks if you don't feel ready to sign on forever. If you're like most people, you'll find your new breakfast provides a great amount of energy and burns longer into the midday without the drop in blood sugar that so commonly occurs with a starchy, sugary breakfast. Once you experience how good you can feel, I think it will be the junk food breakfast that seems strange.

TRANSITION 2: 70/30

This is another giant step: build each meal to be at least 70 percent alkaline (and thus 30 percent acid). Better yet is 80/20, which may be necessary if you are ill. If you're already doing this at breakfast, lunch and dinner will be simple by comparison.

This is a visual measurement, not a measurement by weight or calories. Just give the vegetables the starring role on your plate, where protein (like meat) or carbs (like pasta) might have been before. Make two or three vegetables to go along with what you used to think of as your "main dish." Or make a meal just out of these "sides." Eat a big bowl of salad or vegetable soup with each meal.

The earth is 70 percent water. Our bodies are 70 percent water. Make your plate match: at least 70 percent high-water-content, alkaline food.

TRANSITION 3: RAW

Cooking your food literally takes the life right out of it, and makes it take longer to digest, so the more food you eat raw, the better. Raw foods are alkalizing, and so fit in that 70 to 80 percent we were just talking about. Ideally, all of that three-quarters of your plate is covered with raw, high-water-content food—like having a huge raw salad with a side of brown rice or beans or pasta or tofu. And at least half that portion should be raw. (The other half should still be vegetarian and alkaliz-ing, like cooked soup, or steamed veggies, or stir fry.) Start with that and as you get comfortable with the program work up to the ideal.

This is another reason big salads—and a variety of kinds of salads—are such a great part of this program. Anything sprouted is ideal. With healthful dips and sauces to comple-ment them, as a snack, appetizer, or side dish, raw vegetables are a wonderful, colorful, crunchy, and wholesome way to go. Make sure you include some each time you sit down for a meal.

Raw doesn't mean you have to have all your food cold. It is worthwhile to learn the difference between cooking and simply warming your food. When you do cook, do so as quickly as possible. For example, I (Shelley) am a big fan of quick "steam" fry (like stir frying, except a small amount of liquid is used instead of oil). And when I make a big pot of soup (which is often), I cook it just until it is done—and the veggies are still quite firm. In general, apply heat gently and in moderation. The thing is to not exceed 118°F or 50°C. (The simplest way to check is to stick your finger into whatever you are warming. If you can hold it in without pulling it out right away, you are in the right range. If you have to pull your finger out, the temperature is too high.) Especially avoid the burning, crisping, and browning that can convert otherwise healthy foods into toxins. It is especially important not to heat oils. Steam your food rather than cooking it in oil. (Apply liberal amounts of oil on vegetables *after* warming or steaming.) Or use a nonstick cooking spray made of lecithin. It may take a little experi-mentation to get dishes just right, but the health benefits are worth it. Dehydrating your food is another way of preparing it for additional variety in texture and flavor without cooking it (see Chapter 11).

When the weather is cold, warm cooked food might take a larger portion of your plate (keeping it alkalizing, though), while hot summer months bring more crisp, raw selections. Don't make this program so rigid it becomes a hassle. Keep it flexible and easy, and it will soon be something you do intuitively, rather than something you have to think much about.

TRANSITION 4: DESSERT

Phase out sugary desserts. One sugary dessert can ruin even the best alkaline meal. At our house, we used to be stocked with ice cream and baked goodies like anyone else. First we switched to frozen yogurt and those granola-y "health food" bars, then to Rice Dream® bars. From there we went to simply fresh fruit.

Now that we're fully alkalized, for the most part we don't eat dessert. For us, a treat is a crisp, red bell pepper or thick slices of subtly sweet arrowroot. I realize that might be hard to imagine until you've reached the same place. But tastebuds that may now be dulled by the effects of extreme sugars and salt will come to appreciate the humbler sweetness of vegetables. A biscuit or sweets will seem much too sweet, even intolerable. You will see.

You may have some cravings until your sugar addiction wears off and your blood sugar levels stabilize. Understanding why you get such cravings may help you ride them out. Find other things that will take the edge off, so you won't give in to early temptations. "Cheating" just makes the cravings last longer. However, if you do eat something not on your plan,

waste no time beating yourself up over it. Just get right back
to your plan.

We do break down once in a while, like on holiday, but
only when we are balanced and well. And then we eat dessert
first, or by itself, in between meals, to avoid interfering with
the healthy foods we eat. And we always get right back in the
game.

TRANSITION 5: MEAT

Getting meat out of your diet is painless when you go gradually. Cut back on and then get rid of the red meat
first—beef, pork, lamb, and anything else you have. Make
chicken the next to go, then turkey. Then comes ocean fish.
(You might want to include, as we do, the occasional fish on
your menus. If you do want to have some animal protein
occasionally, I recommend trout or salmon, as they are relatively safe, and are rich in omega-3 oils, which are essential
fatty acids.)

Start with an alkalizing vegetarian meal once a day, then
twice, as you work your way to full-time. At the same time,
experiment with building in more tofu, as well as raw nuts
and seeds, including almonds, hazelnuts, and pecans and sunflower, pumpkin, linseeds, and sesame seeds. Almonds are
especially good—substantially alkalizing and high in protein
and calcium.

Be sure to steer clear of peanuts because of their high fungal
content, however. And in general, avoid rancid nuts and seeds.
If a batch of hulled seeds, such as sunflower or pumpkin seeds,
is sprinkled with broken or sick-looking seeds, don't eat it. It
would be possible to remove the bad ones—in the unlikely

event you had enough time and patience. If you get that rancid taste, an odd bitter sting at the back of the throat, get rid of the batch. Sesame and linseeds, by the way, are almost always okay. Almonds and hazelnuts should either be shelled on the spot or have their brown protective skin intact. Do not use broken, gouged, or chipped nuts.

GO SOAK YOUR . . . NUTS

Soak nuts and seeds to activate their enzymes, eliminate enzyme inhibitors, and partially digest the protein, thus increasing their nutritional and hydrating potential by making all the good stuff they contain readily available to the body. Soaking also makes small seeds such as sesame and flax easier to chew, and therefore to digest.

Place nuts in a container, cover with water to one to two inches above the top of the nuts, and place in the refrigerator for an hour or two or, for almonds, overnight. They will plump up, absorbing the water and the oxygen in the water. Then they will be ready to enjoy. Rinse them off and change the water every day. Keep them totally submerged. Eat them within two days, to prevent molding under the skin of the nuts.

TRANSITION 6: DAIRY

This step may actually be key to the first (breakfast), if you're one of the many, many people I talk to who can't think what to eat in the morning—or to give the kids—if not a bowl of cereal with milk.

The first thing to do is work on milk. Switch to soy milk (though it is hard to find one that isn't full of added sugars in the form of rice syrup) or rice milk (also sugary). Move on to nut or seed "milks." They are good sources of protein and calcium, and have that richness and creaminess that is so pleasing. You can dilute them to taste. They are good for adding texture in salad dressings or soups, or just to drink. (Though when it comes to something to drink, pure, alkalized water and fresh vegetable juices are always your best bet.) I mostly use almond and sesame (aka tahini) milks, and occasionally cook with rice milk or coconut milk. You can make your own (see page 371), if it is difficult to find a supplier.

After you've eliminated milk, other dairy products, such as cheese, yogurt, and ice cream, will be easy to cut out, finding transitional substitutes at first and eventually going without.

EATING OUT

Some people worry that changing the way they eat will mean an end to socializing in restaurants. While it may be true that you'll have another transition to make here, in the restaurants you choose or the dishes you order there, you are by no means stuck cooking for yourself in your own home forever after. Many places have vegetarian restaurants now (there are even a few featuring raw foods!), and more and more general restaurants are offering vegetarian and even vegan entrees. If you stick with Asian cuisines, you'll be sure to find vegetarian options. And most restaurants have salads or side dishes you can build a fine meal from if none of the featured entrees suit your needs. For example, you could put together a pretty good meal with a green dinner salad, an order of the vegetable of the day, and a side of beans or rice or baked potato. Of course, any place with a salad bar will do just fine (though you still have to choose carefully at salad bars, which are almost always stocked with acidic salad dressings with vinegars and sugars and plenty of junk food, too). And don't be afraid to make specific requests—we do it all the time, and they are almost always graciously accepted. (Our most common requests are to hold the cheese, skip the bread, or leave out the mushrooms.) Most chefs are happy to do a basic vegetable stir fry if you ask for it.

TRANSITION 7: YEAST

Bread is another tough one for a lot of families, but you must get rid of the yeast. At our house, we went first to yeast-free bread (your health food shop will have some choices), then to rice crackers, then *sprouted* whole wheat tortillas. In the Recipes section, you'll find some yeast-free breads and crackers you can make yourself (see page 373). In addition to simple substitutions, you must also open up your thinking to meals that don't include bread or other yeast products. If you are one of those people who wouldn't know what to have for lunch if it didn't involve a sandwich, or breakfast if it didn't include toast, this may actually be the biggest challenge for you. Focus on what you can have, what is good for you, rather than what you can't have. The Recipes section gives plenty of meal ideas, along with the recipes, to help you on your way.

Get rid of mushrooms too—they are fungi, just like yeast.

TRANSITION 8: WHITE FLOUR

If you've eliminated yeast breads and baked desserts, you've most likely gotten rid of the major source of white flour in your diet. The other big hurdle is usually pasta. Most recipes will work well if you substitute cooked whole grains such as millet, spelt, rice, and buckwheat. For noodles, in our home soba noodles (Japanese buckwheat noodles) are a favorite, and satisfy the need for a chewy, warm food, especially in winter. We also love mugwort and wild yam soba noodles. If we use pasta other than soba, I try to make sure it is made with vegetables and without eggs—and serve it as a side dish, never a main course.

TRANSITION 9: WHITE RICE

Here's a simple one for you: Switch to brown rice. Or alternate, as I do, with white jasmine or basmati rice, which are natural white rices, or wild rice (or combinations). What we're really after here is all refined grains. You need to give the boot to anything that isn't whole grain. Best, as always, are sprouted grains. And remember, cooked grains belong in that 20 to 30 percent of your meal that is acid (except buckwheat and spelt, which are not acidifying). The one starch you should abandon altogether, because of the high sugar and fungal content, is corn.

TRANSITION 10: ADDED SUGAR

Eliminating dessert might have taken care of a lot of this, but now it is time to scrub out the rest of the unnecessary sugars. Check your cereal, bread, and anything you bought prepared. Don't rely on artificial sweeteners, because they all convert to highly toxic acids that can harm the brain. If you need a sweetener to help you transition while your tastebuds adjust, try something natural like chicory root powder, which you can find at the health food shop.

TRANSITION 11: FRUIT

With the exception of the low-sugar fruits I keep talking about—tomato, avocado, lemon, lime—fruits are intense sources of sugar, and must be eliminated if you are ill or have troublesome symptoms. Once you are in balance, you may still want to use it rarely and with care (and properly combined), as a treat. Fruit has nutritional value, but most have just too much sugar to use freely.

TRANSITION 12: CONDIMENTS

Most condiments are full of sugar, salt, or both. Or they contain fermented or acidifying ingredients. Experiment to find your favorite alternatives to ketchup, mustard, vinegar, mayonnaise, barbecue sauce, soy sauce, and so forth.

Your best allies are healthy oils, lemons, garlic, onion, ginger, and spices. You should also get to know Bragg™ Liquid Aminos, which I use where I once used soy sauce (see Resources). The oils I like best are grape seed oil, linseed oil, pumpkin oil, and olive oils. I use a lot of Essential Balance and Udo's Choice®—combinations of healthful oils (see Resources). All oils should be added to food that has already been cooked, since heating oil destroys its vital components. Better yet, use it to make salad dressings. Rather than cooking in oil, steam your food and add oil as you are serving it.

Lemon and lime add freshness and zest to just about any dish, and because of the low sugar content are alkalizing to boot. They also help stop sugar cravings. They are another key ingredient for many salad dressings. I think I put them in just about everything I make, right down to a glass of water. Garlic, onion, and ginger are all naturally antifungal and antiparasitic, not to mention their nice strong flavors, so include plenty of them, as well.

Getting creative with spices is going to be the key to making delicious meals that appeal to your tastebuds. I love Spice Hunter brand spice blends, like The Zip, which contains onion, paprika, chili pepper, cumin, garlic, jalapeño, coriander, cayenne, and oregano (unfortunately not available in the UK). They take a lot of the guesswork out, but leave the subtle, interesting combinations in (see Resources). Experiment!

MARSHA'S STORY

Okay, I'll admit it. I used to eat bacon or sausage and fried or scrambled eggs for breakfast, with toast with jam—or butter so thick I could see my teeth marks in it. Every day. I never gave it a second thought. And it

only went downhill from there through the rest of
the day.

It seemed like I was always sick with a cold or allergies, my skin and the whites of my eyes had a yellowish
cast to them, and my eyesight was getting worse and
worse. I was always tired, though I accepted the ebbing
of my energy as part of the aging process. When I
needed yet another, stronger contact lens prescription
just three months after the last one (again!), it scared me
enough that I decided to commit to changing my diet,
adding more alkalizing foods and concentrated green
powder and lots and lots of pure water.

I started with breakfast. I started using vegetarian
burgers in place of the meat, and switched to soy margarine instead of butter. Then I learned to smash up
tofu to look like scrambled eggs—I like them sprinkled
with spicy seasonings. Then sometimes I'd put avocado
slices on the plate next to the tofu scramble. Then I
went on to having something totally different for me,
such as soup, or avocado and cucumber over raw (not
toasted) buckwheat.

Where I used to drink lowfat milk, I started with
vanilla soy milk, or almond milk. Now I drink fresh
vegetable juices or put green powder in my water, and
use soy and almond milk only occasionally.

Just changing breakfast gave me much more energy.
It also changed my long-standing habit of snacking
starting by 10:00 A.M. With an alkalizing breakfast, I'm
not hungry again until lunchtime. Then, when I always
had meat, poultry, or seafood (just like at dinner), I
started using meat substitutes, and then extra-firm tofu.

Now I mostly have beans or legumes stir fried with broccoli or other vegetables along with a nice big salad. I've traded peanuts and peanut butter for soy nuts or almonds, and soy or almond butter. I use avocado as a base for creamy soups. I switched from vinaigrette on my salads to olive oil and lemon juice with seasonings. My between-meals snacks have become celery, red, yellow, or orange bell peppers, baby carrots, or cucumber, with almond butter or hummus—where I used to have sweets or crisps. I still get to crunch and munch to my heart's content, and over time my huge sugar cravings have left me.

Since replacing meats and dairy three years ago, I haven't had allergies, a cold, a sore throat, or the flu. By flip-flopping, my eating pattern to 80 to 100 percent alkalizing (where it used to be that much acidic!)— gradually, as I could manage it, not overnight—I gained more energy and stamina. I wake up earlier and stay up later, and my energy level is on an even keel throughout the day. The yellow tinge disappeared altogether. I lost—and have kept off—thirty-eight pounds. My eyesight not only stopped deteriorating, it has actually improved. I've had two new prescriptions in three years now—this time *less* strong each time.

Once your transition is complete, and your symptoms (if any) are gone, and you are at a stable, natural, healthy weight, your body will be in appropriate alkaline/acid balance. Although you've done all this work because you are adopting a new way of life, not some short-term diet plan, I do want to note that a healthy, balanced body can withstand a certain degree of "cheating." Not that I'm recommending it, mind you, but I don't want you to think that the occasional sensational, acidifying indulgence will undo all you've accomplished. Though you'll have to be stricter in the beginning, that kind of treat here and there may be no problem at all for a balanced system. (We probably "cheat" once every other month or so.) And if it is a problem, now you know how to fix it. Remember, it can take up to twenty times the amount of alkaline food to neutralize a dose of acidic food, so you don't want to make your body do that often. It is the everyday bombardment of acidic foods that leave a body completely out of balance. Once you've gone through all these steps, and transformed the way you eat, you'll have gotten well. Now, to stay well, of course you've got to continue on the path you've set so far.

Chapter 8

Cleanse

Our bodies, subjected as they have been to the typical Western (acidic!) diet, urgently need cleaning up. Gradually transitioning to an alkaline diet may be the best way to make this a way of life rather than a short-term "diet" Band-Aid. But at some point, if you want to reap all the benefits of this program, you are going to need to clear out the old to make way for the new. You can't just plant new trees and build new houses—first you have to get rid of the sludge. You need to do a Cleanse to rid the body of impurities, normalize digestion and metabolism, and regain alkaline balance.

If you have an immediate, serious health concern, you may want to jump right into the program this way. Transitioning is all well and good if you have the option, but if you are plagued with negative symptoms, more drastic action may be required. Otherwise, if you are taking your time making a transition, think about doing a Cleanse as you near completion of your transition, so that after your Cleanse you can continue directly to a purely alkaline diet. Finally, even if you are planning a longer transition, you can begin with a Cleanse as a way to sort of jump start yourself. Of course, you can do another Cleanse after you're on a pure alkaline diet any time you want or need it. We like to do a springtime Cleanse, just as the earth renews itself after a long winter.

A Cleanse is basically a juice fast—though not actually a fast, which technically involves nothing more than water and requires close medical supervision. I think of the Cleanse as a liquid feast, rather than a fast, as in reality you are getting twenty times more food than you would generally, because you are drinking it in such concentrated form.

The length of a Cleanse will vary from person to person, depending upon one's current situation and how one tolerates the Cleanse. Typically, three to ten days will give you the results you want.

As you would with any fast or restricted diet, you should consult your health care practitioner before beginning a Cleanse.

THE CLEANSE

The Cleanse eliminates acid wastes and negative microforms throughout your body, detoxifying your blood, tissues, and digestive system. You have to get rid of the pollution that's built up in your body, and especially the colon, from eating all those poorly combined foods, processed foods, fried and overcooked foods, simple starches, and sugars.

We recommend at least a two- to three-day Cleanse

(liquid feast), and up to ten days if seriously ill (though you should seek supervision for an extended Cleanse). Seven days is good for everyone who can manage it. If you are not facing any particular health challenges and can't take much time out from your schedule, just forty-eight hours can still be beneficial—and can be done over a weekend (and better yet, a long weekend). The shorter fast is also good for older people or children and teens. You can also use a brief Cleanse to get yourself going after you've "fallen off the wagon," or just to give your body a periodic break. Once you are fully transitioned, I recommend a liquid Cleanse at least one day a month—twenty-four hours will do—just to give your body a break from solid food. However long you do it, a Cleanse gives your body a good rest.

While you are on a Cleanse, you should drink at least four liters/eight pints of purified water, with hydrogen peroxide or Aerobic Oxygen drops, a day (with lemon or lime juice if you like). You can include, as part of or in addition to those four liters, six to 12 eight-fluid-ounce or 250 ml. glasses of fresh green vegetable juice, to help clear the toxins out of your system and make your body more alkaline.

Try juicing cucumber, kale, broccoli, celery, lettuce, collards, okra, wheat grass, barley grass, watercress, parsley, cabbage, spinach, alfalfa sprouts—just about any other green vegetable that appeals to you. The Recipes section has many combinations to try, but here's one that's all green: one cucumber, one stalk celery, one-third bunch parsley, a handful of alfalfa sprouts, and some spinach leaves.

When you're juicing in general, you can use carrot or beet to make the taste milder. That works because they contain sugar, which gets concentrated in the juicing process, so you always want to use them moderately. During a Cleanse, however, if you are dealing with an acute condition, you don't want to use them at all.

Some people, however, find plain green juices difficult to stomach in the beginning. If that's you, you might try "backing into" a Cleanse by starting out for a few days using carrot and beetroot, just to get used to drinking fresh vegetable juices, then weaning off the carrot and beetroot gradually, and upping the amount of greens. (If you do choose to use it, take it easy on the beetroot at first—maybe half a medium-sized beetroot a day to begin with—as it can add considerably to the bowel-cleansing effect of the juice.)

Make sure you always dilute the juice with water (whether or not you are on a Cleanse)—ten times as much water as juice—and add four drops of Aerobic Oxygen per cup. That will increase the alkalinity of your juice from 6.2 to 9.5.

If you cannot make fresh juice, you can use the concentrated green powder mixed into water instead—a quarter of a teaspoon per eight fluid ounces/250 ml. of water (plus four drops of Aerobic Oxygen). You can add the green powder to any or all of your four liters of water whether or not you are juicing. You can also take one to two capsules of green powder with your fluids and juices if you need the added convenience.

Instead of juice, particularly at your usual mealtime, you may prefer raw, puréed soup, like Popeye Soup (see page 274). You can also drink broth from soups like Healing Soup (see page 275), and teas such as Essiac™ (a special blend of herbs), and raspberry leaf. Add three tablespoons of essential oils (cold-pressed linseed, virgin olive, borage, primrose—or a blend, like Udo's) to your soups, or take it by the spoonful.

Although many people have no trouble with the Cleanse, it isn't always easy. Naturally you may feel some hunger during the Cleanse. It is really just those greedy microforms screaming to be fed, so resist any urge to break your regimen unless you find you can't keep up with your daily routines. The initial hunger pangs are the worst, but once you are over the hump—usually, by the third day—you may actually experience an upsurge in energy and feel no need for solid food.

When hunger does strike, mineral supplements, particularly chromium, may help (see sections on supplements, pages 159 to 163). Also, try drinking lots of water, or water with a quarter of a teaspoon concentrated green powder per cup, to take the edge off. If you must resort to solid food, eat fresh, raw veggies, or Healing Soup (see page 275) with the optional veggies added.

BASIC SUPPLEMENTS

There's a whole chapter on supplements coming up, so here I'm just going to mention the ones that are particularly useful during a Cleanse. Check out Chapter 10 for details about the individual supplements.

Supplements can maximize the effects of a Cleanse and bring the body into balance more quickly, controlling negative microforms and excess acidity. In general, supplements should be taken with meals or drinks—one capsule three times a day, with meals. Colloidal supplements should be taken under the tongue, three to five drops three times a day, away from meals. If you have been struggling with chronic or severe health problems, use up to six to nine times a day.

Where the directions are different for a particular supplement, I've noted this in the description. You could also follow directions on the package, but remember that less, more often, is better than more, less often. That is, taking one capsule six times a day is better than two capsules three times a day or three capsules two times a day.

During a Cleanse, I recommend you take several things. The two most crucial are **pH drops** (Aerobic Oxygen/hydrogen peroxide) and **concentrated green powder**. Add four drops of Aerobic Oxygen/hydrogen peroxide per cup of pure water into all your pure water. Mix a quarter of a teaspoon of the concentrated green powder into eight fluid ounces/250 ml. of pure water three times a day, or take one capsule three times a day with a "meal" or drink. If you do nothing else in the way of supplements, do these.

PETE'S STORY

I was diagnosed with bladder cancer over three years ago, stage three (of a possible four)—not good. The doctors were worried that the malignant tumor could have spread to nearby lymph nodes and might have extruded through the bladder wall. It had definitely blocked one of the ureters connecting the kidney and the bladder. I underwent two months of chemotherapy, but I had such a severe negative response (my wife thought I was going to die) that my doctors recommended discontinuing treatment. Unfortunately, as much as I had made it through had had no effect on the tumor. My doctors recommended removing the entire bladder surgically.

I was determined to fight the cancer, but I just couldn't believe radical surgery was my only choice. I set out to find out about my alternatives, and my search led me to Dr. Young. My blood turned out to be highly acidic, and live analysis of my blood showed my body chemistry to have a high degree of toxicity. My blood cells were a mess from years of eating garbage and ignoring my health.

I did a ten-day fast, with vegetable juices and soups, which, to my surprise, wasn't that bad. I started taking the recommended supplements, especially the pH drops and the concentrated green powder. After the ten days, I kept to vegetable meals, following the program to the letter. This was a radical change from my former diet, but I was determined to beat the cancer.

I told my doctors about my new approach, and although they were skeptical, they had nothing else besides drastic surgery to offer me, and agreed to monitor the tumor with intermittent MRIs. Many of my family and friends thought I was crazy. I had many tearful discussions with my wife, who was very supportive throughout, about which path to follow. One of my best friends accused me of trying to kill myself. But I remained steadfast. I would not vary my commitment to see this program through.

The first two months I lost between fifteen and twenty pounds, which confirmed my loved ones' suspicions, as I hadn't been overweight to begin with. But I felt better each week, with more energy and clearer thinking. I knew I was doing the right thing even before repeat blood analysis showed considerable improvement.

By three months into the program, there was no signs of any cancer spread, but my doctors kept suggesting surgical removal of my bladder. I was very opposed to that, naturally, especially now that I felt sure this program was working. I did agree to a diagnostic procedure, which revealed the tumor had not only shrunk, but also was suspended on a stalk to the bladder wall, no longer fully attached. The opening of the ureter was clear, and they were able to remove all of the old tumor during this procedure. The doctors took sections of muscle tissue of the bladder to test them for pathology (clear) and examined the ureter all the way to the kidney (also clear). There was no cancer in my bladder whatsoever, and only remnants of the carcinoma in the degenerated (and removed) tumor. I had won!

Amazingly, the surgeon *still* recommended removal of the bladder for what he called a "cure." I continue to get intermittent MRIs to monitor my bladder, but I know as long as I stay alkaline, the cancer will not recur. I just told the surgeon, thanks, but no thanks.

A close second in importance are a **multivitamin** and a **multimineral,** with cell salts. Each capsule of the broad-spectrum vitamin should contain at least 500 milligrams and the multimineral 500 milligrams. Among other things, nutritional deficiencies increase the toxicity of mycotoxins, so you want to be sure you get all that your body needs. The minerals are particularly important because all other nutrients, including vitamins, proteins, enzymes, amino acids, and carbohydrates, require minerals for normal biochemical functioning.

Finally, I'd also recommend **chlorophyll**; a **noni fruit concentrate** for its antifungal and antiparasitic properties and enzyme activation; colloidal **silver, rhodium**, and **iridium**; and an antimycotoxin formula, ideally combining **n-acetyl cysteine, l-taurine**, and organic **sulphur**.

BOB'S STORY

I was lining a water storage shed with highly flammable materials, on the roof doing fire watch while my partner was inside working. The pump attached to the truck was grounded improperly, causing static electricity in the lining of the hose, which combusted in a large explosion. My partner was burned over 90 percent of his body, and died in the hospital four hours later.

The flames shot up through the skylight, through which I was gazing, so I received a flash burn. I was blown off the roof and landed on the ground, my face and hands badly burned. I was taken by air ambulance to the hospital burn unit.

My head was swollen to twice its normal size, my eyes swollen shut, my face black and crisp. My nose had more or less disappeared in the swelling, so breathing tubes had to be inserted into my nasal cavities. My ears were severely burned, my fingertips charred and numb, my fingernails melted.

I was told I had second- and third-degree burns on my face and third-degree burns on my hands. The head of the burn unit told me I would be in the hospital for two or three months and would need many skin grafts, especially to my hands. He said he thought I would lose my fingertips and my ears.

The very next day, my mother brought me colloidal silver, with directions to use it externally and internally as frequently as possible, because it assists in growth of new tissue. She had learned from Dr. Young that silver's negative electrical charge counters the positive charge of a damaged body area, bringing it back into balance and enabling the body to regenerate and heal itself. Mom sprayed undiluted colloidal silver onto the burned areas, and I took it by drops under my tongue. I also took cat's claw, germanium, linseed oil, and vitamin C, as Dr. Young recommended, and, once I was well enough to eat, lots and lots of greens. As soon as I was off intravenous feeding, my mother brought me wheat grass juice, green juices, and concentrated green powder to bolster my nutrition and healing.

My mother applied the silver many times a day. It was absorbed into my skin instantly. It felt cool and tingling and loosened the tension on my face, hands, and fingertips. The tips of my fingers and the skin under my nails started throbbing and tingling. The staff said it was because my nerves were healing and the blood was circulating.

After only one day of treatment with silver, it was obvious that healing was happening very rapidly. New tissue and skin grew back at an accelerated pace. The swelling of my head diminished rapidly and the breathing apparatus came off almost immediately. The plastic surgeon told me I was healing twice as fast as any burn patient he'd ever seen in his long career.

When I told him it had to be because of what my mother was giving me, and the alkaline diet, the doctor asked to see what I was using. He read the ingredients and said he didn't see anything wrong with using it (in fact, the healing salve used by the burn unit contained silver), though he noted that the hospital could not be held responsible if anything negative occurred, since it was not a hospital-prescribed medicine.

I still have my fingertips and my ears. I had one skin graft on each hand, but my own skin grew back so fast with the help of the silver that the grafts were useless and actually fell off! I was out of the hospital in two and a half weeks—not two or three months! What most amazed me was that my new skin looked better (smoother) after it healed than it had before I was burned.

The burn unit staff marveled that they had never seen anything like my recovery. The nurses asked if they could use my pictures to show to other burn patients, to help explain the healing process. I was happy to say yes, but asked them to tell them about the colloidal silver, too, and the greens I was getting. They said they'd do their best.

I know I am truly blessed to have healed from these terrible burns with no damage—and no scars.

A TYPICAL DAY ON THE CLEANSE

All the details may at first seem daunting, though you'll quickly get used to the routine. In the meantime, to help you sort things out, here's a typical schedule, which you can modify to suit yourself.

7:00 A.M.: 1 liter/2 pints pure water with pH drops (with lemon or lime juice if you like)

7:30 A.M.: colloidal supplements

8:00 A.M.: juiced greens and capsule supplements

9:00–12:00 noon: 1.5 liters/3 pints pure water with concentrated green powder and pH drops

12:30 P.M.: colloidal supplements

1:00 P.M.: raw soup or juiced greens and capsule supplements

2:00–5:00 P.M.: 1.5 liters/3 pints pure water with concentrated green powder and pH drops

5:30 P.M.: colloidal supplements

6:00 P.M.: soup or juiced greens and capsule supplements

7:00–9:00 P.M.: pure water with pH drops as desired (with lemon or lime juice if you like)

WHEN WE SAY CLEANSE, WE MEAN *CLEANSE*

All this can have quite a laxative effect. Green juice alone can do it. In addition, that is the way your body physically gets rid of the bad stuff. It doesn't just evaporate. This is just what you want, to make sure you get rid of the pollution that has built up in your body, and in particular in your intestines and colon. Until you know your body's response to the program, it is wise not to have anything else planned so you can focus on the Cleanse—and just stay fairly close to bathroom facilities. Be prepared to pay a visit at least six to ten times a day as the toxins clear from your body.

If what we have described so far *doesn't* have this effect, you'll want to add **a mild, natural laxative formula** to your program. Look for one including butternut root bark, cascara sagrada bark, rhubarb root, ginger root, licorice root, Irish moss, and cayenne. Take four capsules every four hours.

Aloe vera juice is another cleansing aid. It helps break up pockets of protein, especially in the small intestine. Add one tablespoon of cold-pressed, whole-leaf juice to your green juice, or take with a "meal."

Remember, these products are designed to work! Don't be caught off guard, and by all means, modify your approach if the results seem too vigorous.

For serious cleansing, acute or difficult conditions such as chronic constipation, diverticulitis, Crohn's, irritable bowel syndrome, or chronic diarrhea, you may want to add (or use instead) a **lower bowel cleansing formula or other intestinal cleanser.** Look for an herbal mixture containing cascara sagrada (more than in the mild laxative), turkey rhubarb root, psyllium seed, barberry root, ginger root, fennel seeds, red raspberry leaves, and cayenne. Take four capsules every four hours during the Cleanse. Adults over 60 should take one to two capsules every eight hours. If you are having trouble eliminating, this lower-bowel-cleansing formula will get things going!

WHAT TO WATCH OUT FOR

During a Cleanse, toxins are dumped from where they've been stored in the tissues into the blood so they can be eliminated. That means that for a while, your blood is actually dirtier than it started out. That can mean you feel worse before you feel better. Different people experience varying degrees of unpleasantness, or none at all, during this "healing crisis," which can include nausea, weakness, dizziness, headaches, lightheadedness, rashes, bad breath, flulike feelings, and fatigue. (A note on rashes or other skin reactions: Do not suppress them with medications. At most, use a pure moisturizer or liquid vitamin E.)

When and if this happens, lots of water with pH drops and fresh lemon or lime juice will help flush the toxins (and their negative effects) out quickly. If you are having symptoms, increase your daily water intake.

A healing crisis is actually a good sign. But it can be too intense, and therefore discouraging or even harmful. So monitor your progress closely. Some mild discomfort can be expected, but you should not experience undue discomfort. A healing crisis should be short-lived. If you experience an intense healing crisis, spread out your dosages of all supplements more. If it does not subside within twenty-four hours, seek assistance to find a way to Cleanse that will work for you. Anyone with serious health problems should seek some professional supervision before beginning this program, including the Cleanse. With guidance, you should be able to minimize or even avoid a healing crisis.

There is no medal for who can Cleanse for the longest period of time, so do not push your body too hard for the sake of reaching a specific number of days. You should feel successful however many days you are able to Cleanse. Each day you remain on it is one more day ridding the body of acid buildup. Each day after brings you closer to a fully integrated alkaline lifestyle.

OKAY, I'M CLEANSED. NOW WHAT?

You've removed the stockpile of debris from your body, clearing the way for optimal health. The next step, after taking away what your body *doesn't* need, is to provide what it *does* need. If you give it the vital materials it needs to construct new and healthy cells, your body will heal itself and be restored to balance and harmony. This second phase should run for at least five weeks, and ideally seven weeks, after your Cleanse is over (for a total of six to eight weeks, so add in a few more days if your Cleanse isn't an entire week).

To this end, no surprise, proper diet is essential. The key to this phase, as you go back on solid food, is reintroducing a still-limited range of healthy foods, as well as keeping clear of certain foods.

You want to keep your diet extremely low in carbohydrates, focusing on dark green and yellow vegetables; sprouted soy beans, seeds, and grains; nuts and essential fatty acids. At least 40 percent of it should be raw. The higher the level of enzymes in your food, the faster you will repair and rebuild your body—and cooking destroys your food's natural enzymes and life force.

Continue with daily green juices, but add some variety. Drinking juice just before a solid meal is good for digestion, but you should also have juice on its own. Beyond that, for this phase you want to avoid (in addition to the unhealthy foods described in earlier chapters) all complex carbohydrates, including high-carbohydrate vegetables (potatoes, sweet potatoes, peas, winter squash), all grains, and starchy legumes (meaning all of them, except soybeans and lentils), as well as sugar fruit.

Otherwise, you want to eat as laid out earlier in this book, focusing on vegetables, especially green and yellow vegetables, and including healthy oils, low-sugar fruits (tomato, avocado, lemon, and lime), soybeans, tofu, lentils, and raw seeds and nuts (of the healthy types), preferably soaked.

If your Cleanse was shorter than seven days, or included some solid foods, you might want to proceed with even more than 80 percent of your diet being alkaline for these six to eight weeks.

During this phase, the principles of proper food combining are crucial (see Chapter 6).

ADDITIONAL SUPPLEMENTS

At this point you should continue with the supplements already begun and add a few more. Again, you'll find more information on supplements in Chapter 10, so here I'm only going to mention the ones that are particularly helpful in this phase. The two most generally applicable are **digestive enzymes**, including **amylase**, **protease**, **lipase**, **papain**, and **bromelain**; and an **antiyeast formula** containing **caprylic acid** or **undecylenic acid**. Beyond that, depending on your situation, the following *optional* supplements can be necessary or valuable: a **probiotic** mix of **acidophilus** and **bifidus**, and an (if you can find one in the UK) **antiparasite** formula containing **black walnut hulls.**

MOVING FORWARD

Congratulations! Now you're ready to move on to the full program, assuming you've made good progress to this point. Now you can add one serving per meal, properly combined and not to exceed 20 percent of your diet, of the following foods: starchy vegetables (peas, potatoes, winter squash, sweet potato), legumes, and organic, unstored whole grains (millet, spelt, buckwheat, kamut, quinoa, brown rice, and wheat). And that's it. You're alkaline.

As you add foods after a Cleanse, carefully observe your body's reactions. If any symptoms return, a longer period of totally alkaline meals may be in order, until more healing takes place.

At this point you can add additional supplements to address specific symptoms (if any remain). You'll read more about how to do that in Chapter 10.

SECOND NATURE

All this may seem a little overwhelming at first. That's normal when one is faced with a lot of new information. Don't be discouraged. Take some time to familiarize yourself with these concepts. Don't be intimidated. Remember how complicated driving a car was in the beginning? Now it is so much a part of you that, if you are like most people, you can think about many other things while doing it. The same will happen as you gain experience with your new lifestyle.

The combination of cleansing when necessary or beneficial and a good diet is designed to keep your body in balance over the long term. It will restore pH balance, stop overgrowth of negative microforms, and heal the damage resulting from the toxins they emit. I do want to note, however, that as far as this biochemical approach can take you toward wellness, you cannot overlook the environmental, intellectual, psychological, emotional, and spiritual factors that also influence your overall well-being. To truly achieve optimal health, you also need to break the pattern of negativity that feeds sickness in disease. The acidic diet you're leaving behind is just one example. You also have to deal with chemical exposure, prescription or recreational drugs (including cigarettes and their nicotine), and, less obviously, negative thoughts, words, and deeds.

The more closely you adhere to the guidelines in this chapter—and in this book—the better your results will be. You should begin to experience success quickly, so I want to caution you against believing that just because you are better, you are truly well. Especially for those who have been struggling with health challenges, this may be unfamiliar territory. Relief of your symptoms is all well and good, but persist with the program and into a completely alkaline way of eating, and you'll experience total wellness that might be beyond what you can imagine while you are plagued by symptoms.

Healthy living should be instinctive, but clearly somewhere along the way humans have lost the ability to sense it. At this point, extraordinary means are required for its recovery. Fortunately, you hold those means in your hands right now. But what we all must do is get our health under control—then get past it. Don't let it become an obsession. Don't become its slave. A healthy body and mind put us in a position to better serve our universe. So they are worthwhile aims, certainly. Just don't get so absorbed in attaining them that you miss out on the unpredictable play of Life!

Chapter 9

Motivation

These, then, are the practical steps involved in switching your body over to this way of eating. Breaking it down this way, taking it slowly rather than making an overnight switch, is all many people need to get started. They get the why, then the how . . . and they are on their way.

But there is an important internal process that happens alongside the practical preparations, and it can be a stumbling block for some people. The people who are already off and running probably went through it without even being conscious of it. For the rest of us, here's a quick look at what has to happen to allow for a change like this one.

Here's the whole process: awareness, investigation, knowledge, potential, motivation, action, patience, results. You have to start with *awareness*. Maybe you looked in the mirror one day and realized, "Whoa! I'm getting fat!" Or you recognized that you just don't feel as good as you used to. Or you decided once and for all that you were sick and tired of being sick and tired.

That awareness pushes you into *investigation*. You saw your doctor, called a diet center, bought this book. You started to look into how and why you got into the situation you are in, and how you can improve it or get out.

That is, you develop *knowledge*. In this case, you learned about acids and bases, yeasts and fungus, mucus and mold, and what it all does to your body—and the options for preventing all that damage and restoring optimal health. Maybe you saw it work in someone else's life. In short, you identify the *potential*. That inspires *motivation*, which moves us into *action*, and, with *patience*, leads to the *results* we desire.

JO'S STORY

I spent a lot of time and energy seeking the perfect physique. I was obsessed with working out and running. I ate a strictly low-fat diet. Now I know that I was practically living on sugar, but back then I felt that the carbohydrates I was consuming were the only way to get the energy I needed for my thirteen- to fourteen-mile runs. But as time went on, instead of increased energy, I found that fatigue, irritability, and depression had become a part of my life.

In a way, I wasn't surprised, as I'd watched several people in my family struggle with mental disturbances. Part of what I witnessed was how they deteriorated as a result of the drugs they took that were supposed to help them, so I refused drug treatment myself. But I didn't want to keep on living the way I was.

Fortunately, that's when I found The pH Miracle program. I came to realize my fatigue and depression were just expressions of the way I was eating, living, and thinking, which was upsetting my body's chemistry. When I cleaned up my blood through diet and lifestyle modifications, everything changed. I no longer suffered from depression and mood swings. My whole attitude became positive as I reclaimed the vibrancy and passion for life I once had as a young person. Talk about motivation!

Still, like most people I have my moments of weakness, when thoughts of doughnuts and bagels dance in my head. I occasionally indulge in whatever my tastebuds ask for. The big difference is, now it is because I just want something different—not because I'm desperate for a sugar fix! Anyway, sweets just don't taste as good anymore. I'm more likely to want vegetables and green drink now, and my yearnings for junk food are few and far between.

When I think back to how sick and tired I was before this program, it is easy to stay motivated. And then there's the way my husband looks at me in a tight pair of jeans! Among other things, I know he's motivated to stay healthy too. It's a matter of choice—quite a simple choice, actually, once you fully understand it, simple to make and simple to keep.

You might be anywhere along this spectrum right now. If you find you are stuck, take a look to identify which phase it is, exactly, where you break down. Maybe you're motivated, but haven't organized to take action. Maybe you took action, but lost patience and never saw the results that would have come. But there are no magic bullets or instant solutions! Maybe you investigated, but didn't really "get" the answer—

you haven't truly developed knowledge. Maybe you are waiting for knowledge to somehow drop into your lap out of the
clear blue sky, and haven't undertaken any investigation.
Maybe you haven't even had that initial realization that something must change, and you're waiting for that awareness.

Wherever it is you find yourself stuck, pick it up from there
and follow through to the end. Conscious effort is usually all
that is necessary to jump start the process. And remember, as
you go through this process, any time you have trouble you
can come right back here to find where you are stuck. Maybe
you will do great eliminating meat and dairy and added sugar,
but just can't let go of the fruit. Even for that kind of detail,
this process will help you work it out. You should expect there
to be more than one hurdle you have to clear to win the race!

Some people will take these steps out of order. The most
common variation is to start with the action, even if you haven't
developed the knowledge or really don't believe the potential.
Try this program out. This is more or less a trial-and-error
approach, or a wait-and-see attitude. For some people this is
the most persuasive of all. If you try the program, and you start
to get results, *then* you truly recognize the potential, experience
the knowledge, and recommit to investigation and action, with
renewed patience, and, in the end, even better results.

So if you're one of those I haven't yet convinced, consider
taking a chance, trying it for yourself. You may be the only
one who can convince yourself. No one's making you take a
loyalty oath for a lifetime on the program as a prerequisite. So
go ahead. You've got nothing to lose by trying—and everything to gain.

Motivation is so important—and sometimes so difficult—that
I want to devote this chapter to exploring it. You must always

stay focused on your motivation—and the results you want to see happen—as you begin to make these important and healing changes. If you forget why you want to make the change, the risk is that you'll start to view the program as a kind of deprivation, rather than the gift of life and health it is. You have to be specific about your motivation to make this program a new way of life, not just a temporary patch. That focus will be helpful when you crave foods that aren't good for you, or when well-meaning friends or family want you to indulge with them. If you stay focused on your own motivation, and gently but firmly turn down foods that could contribute to your condition, those around you will eventually understand and respect your choices. (And when they see your results, don't be surprised if they want to know how to do it themselves!)

Starting on this program is making a commitment to yourself, to your life, your health, your well-being. I recommend a five-step approach to making sure you honor that commitment—and honor yourself.

1. **Define and record your motivation.** The first step in focusing on your motivation is to state the change you want to see in your life. Identify it as specifically as you can, in the form of a desired outcome: "I want to be free of my headaches so I can enjoy more activities with my family"; or "I want to be slim"; or "I want to have the energy to enjoy life, my family, and my work"; or "I want to be really well." Someone with acute illness might simply say, "I want to live!" Even someone already healthy has a motivation, perhaps, "I want to prevent and avoid illness and enjoy a long life." Once you are clear on what your motivation is, *write it down.* Seeing it in print makes it somehow more real or more serious.

2. **Set a realistic and appropriate plan.** In order to make change that is effective and lasting, we must be honest with ourselves. Seriously consider exactly what will have to be done to make your motivational idea a reality, taking into account just where you are starting from, and plan the specific steps you need to take to get you there. Make sure to keep it reality-based. If you plan too much too soon, you'll defeat yourself before you are even out of the starting gate. If you oversimplify, giving in to wishful thinking, you'll never get results.

   Include a time frame in your plan, but remember the blue ribbon doesn't necessarily go to the first one over the finish line. Pace is vital. This is more like a marathon than a sprint.

   Break each new behavior change down into steps. This program should be an exciting process, not an overwhelming, overnight challenge.

   The more imbalanced or sick you are, the more reason you have to find motivation, and the more commitment is required to find balance and wholeness. If you are very far from where you live, you won't be able to get home by going only halfway back. If you've got garbage in your house that is attracting flies, you can't get rid of the flies by taking out half the garbage. In this vein, you're not going to shed a long-standing health concern by planning to go off coffee or sweets, but staying on bread and fruit (though you might indeed see *some* improvement— you just shouldn't confuse that with really getting well). Unless you are acutely ill, you want to avoid extremes.

   A more realistic plan would go something like this: "I will take the next six months to do all I can to eliminate my headaches. I will make two major transitions each month, until I've made all the necessary changes to bring my body back into a state of balance." You would then spell out the specifics appropriate to you and your situation, such as, "I will start drinking fresh vegetable juices and try an alkaline meal at least three times a week," and, "I will eliminate all added sugars for one week, then phase out the high-sugar fruits until I reach a state of balance," and, "I will drink four liters of properly prepared water a day, including one mixed with green powder."

This book lays out the basic steps, in a reasonable order. But individuals will work it in their own way and must take personal responsibility for themselves. You are unique, and the most anyone else's program can do is give you a good foundation. You have to build your own house. This is your chance to customize the design to make sure you are comfortable in your new home—and to maximize your success.

Write your entire plan down, too, and keep it where you can see it.

3. **Practise your new habits.** Whatever your plan is, start to *do* it. Practise having a totally alkaline breakfast at least three times a week, then five, then seven. Practise going without a heavy, sugary dessert after dinner. Practise skipping the alcoholic beverage before or with dinner. Practise choosing a vegetarian entrée in the restaurant.

Practise is repetition, and repetition of any skill increases your ability in that skill. As you practise these new, healthy habits, you will develop a skilled, intuitive way of feeding your body what it needs without feelings of deprivation. Instead of thinking about what you can't have, you will feel blessed with the abundance available to you.

4. **Evaluate, review, and reward your progress.** Monitor your symptoms. Are they diminishing, or have they gone? Do you feel lighter, brighter, more energetic? Have you lost some weight? Do your clothes fit better? Ask yourself if you stayed with your plan, and if the plan is still the appropriate one for you. If you're having trouble making the change, re-evaluate your goal. If it seems too harsh—or too lax—adjust it (within the bounds of appropriateness to your condition) so that it is more reasonable for you to achieve. And to the extent you did what you set out to do, reward yourself. Actually the best reward for bringing your body into balance will probably be simply the disappearance of symptoms. If you desire a reward beyond that, be careful to avoid the common tendency to make food a reward—especially unhealthy food. Buy yourself a new pair of trousers for your new size. Make time in your schedule to do something you love but rarely get to do—read a book, take a bubble bath, listen to a favorite piece of music. Treat yourself to a class that's always intrigued you. Hang new art in your home. Choose something that nourishes your self as your new diet nourishes your body.

5. **If you "fall off the wagon," simply get back on again.** Feeling guilty or down about your mistake is a waste of time and a drain of energy. Forgive yourself and just start moving forward again. Restate your motivation and goals.

If you find yourself having a very hard time staying on the program, and giving up too easily or too often, you may be running into the fact that eating is, for most of us, not just physical—it is also highly emotional. Many times, we use food for comfort in response to stress, when what we really need is to find a way to relieve the stress permanently. We need to understand why we turn to food for comfort, and where else we might get that comfort. If exploring that is more than you can do alone, or if you feel deep emotional wounds, I encourage you to seek a good counselor to help you shed some light on the dark areas.

For this program to work for you, there must be a change in the way you eat and live. Perhaps most important, there may need to be a change in the way you *think*. There are a handful of commonly harbored—but dangerous and self-destructive—attitudes that prevent people from finding the wholeness and healing this program provides. I want to review them briefly here, so that if one of them rings a bell for you, you can take a second look at your stance.

This condition runs in my family so there's nothing I can do about it. This attitude, like Michelle's (see page 182), prevents you from taking personal responsibility for your own body and your own health, and therefore from taking steps to control, eliminate, and prevent the disease. Certain apparent weaknesses or a predisposition to symptoms may be inherited, but there is always much you can do to foil fate. Avoiding this sort of resignation is the first step. Genes never tell the whole story (and usually, you don't even know if you have the genes in question anyway). Even if you do have the genes, there's always something you can do to lessen, put off, or curb a condition.

My doctors say they've done all they can. Michelle had to overcome this one, too (see box). What she came to realize is that no matter how wonderful your doctors, if you are still struggling with symptoms, not everything has been done.

But the more important point here is that the ultimate responsibility for your health and your body lies with *you*, not with any doctor. And you are the best judge of what is working and what is not, and what feels right trying and what doesn't.

MICHELLE'S STORY

Besides working full-time and being the mother of three great kids, I've been something of a professional patient for my entire adult life. It started very young, with chronic upper respiratory problems two to three times a year until the age of ten. From there the list just goes on: two serious bouts of pneumonia, chronic gastrointestinal problems, menstrual disorders, serious concentration and information-retention problems, adrenal/thyroid problems, fatigue, dizziness, hiatus hernia, anemia, hypoglycemia, muscular skeletal disorder, pituitary dysfunction, severe nervous tension, on and off addiction to prescription amphetamines (doctor's orders), an eating disorder that required hospitalization, and depression.

I've taken two hundred prescription medications—not to mention what I tried over the counter—but nothing worked. Now I know why. But at the time it was all I could do to find the strength to go on and act as if everything was fine.

So maybe you can understand why it almost seemed like good news when my doctor suggested a total hysterectomy when I started to suffer severe ovarian dysfunction. "Fine," I said. My mom died young of ovarian cancer anyway. I thought the hysterectomy might solve all my health problems. Maybe they all stemmed from having a bad hormonal imbalance.

After the surgery my doctor came in and said he thought he'd found the reason I'd suffered all the abdominal discomfort—my uterus was the size of a large cantaloupe. "Great!" I thought—with the cantaloupe out, surely my troubles would end.

And my abdomen did feel better. But that was about it. I still had an array of the same old problems. One doctor told me that women tend to get depressed and suffer gastrointestinal problems and that I should just learn to live with it. He offered me hormone replacement (which only made me feel worse) and noticed some Candida around my nails for which he prescribed Diflucan.

An allergist told me I was antibody-deficient, and referred me to another doctor who put me on more Diflucan. It didn't help much, and I worried about the potential for liver damage.

I found a doctor who had me try all kinds of new medicines, with no luck. Then he recommended surgery to remove a portion of my large intestine. What a catastrophe that turned out to be! I had so many complications, infections, and antibiotics my planned four-day hospital stay stretched into seventeen. Need I say that as soon as I got out of the hospital and started really eating again, all my symptoms returned?

I went back to pretending everything was fine, with the help of occasional amphetamine prescriptions to keep me going. But I knew this was all wrong. You shouldn't need drugs of any kind on a daily basis. And you certainly shouldn't feel rotten all the time (with or without the drugs). I developed two new symptoms: My hearing was becoming muffled and I was losing the vision in one eye. I lost it completely one night after eating a piece of cake! When my bad days really started to outnumber the good, I made a New Year's Resolution to get myself *well*.

My first real clues came from a book I found in the health food shop. I started doing my own research, but when I looked for a doctor to work with me I was told over and over again that this was a bunch of loony hogwash.

Then my cousin told me about Dr. Young. As I learned about The pH Miracle program, right away I felt he was talking about me and my problems. I knew, finally, that this wasn't all in my mind—and that there *was* a solution. I was also glad to realize I wasn't alone. You can't imagine the burden that was lifted from me that day.

Within a week of starting on the program, my energy picked up. My digestive system became more comfortable. The most major change was the clearing of my "brain fog." I gained alertness, and my concentration level was 100 percent better. (I was able to sit down and write this!) My sore, bloodshot eyes completely cleared up. My skin, hair, and nails were glowingly healthy. I can't remember a time when I actually felt entirely well like this, but now I am ready for a long future of it.

One way or the other, I'm going to die when it's my time, so I might as well enjoy myself and eat anything I want in the meantime. Whether or not you believe your days are numbered by some higher authority, you have the power to fill whatever time you have with misery thanks to careless abuse of your body—or to enjoy every last minute you have in excellent health. Death may not be something to fear, but the greatly diminished quality of life that usually precedes it *is*.

I'll just pray for healing. God will heal me. There is nothing to contest that faith and prayer can heal us, and I do believe in miracles. But to ask for healing while refusing to cooperate in every way we can to heal ourselves is, to my mind, a futile attempt at becoming whole. Also, requesting wholeness while knowingly facilitating the disease process sounds like, "Help me, even though I won't help myself."

The value of change is the result. In this case, the peace and harmony that can arise from embracing the change of this program as part of everyday life are yours for the asking. You've made it this far because you want the best for yourself and those you love, and by following these steps you'll be whole and strong, and able to enjoy the best this world has to offer. If that's not motivation, I don't know what is.

Chapter 10

Supplements

Even if your diet were ideal, we'd still recommend supplements. Today's food is a mere nutritional shadow of what it once was—and should be. It is grown with artificial fertilizers in depleted soil and sprayed with pesticides to within an inch of its life. It is harvested early and shipped long distances, and it languishes in trucks and warehouses and on shelves, losing nutrients with every passing minute, for far too long before it ever reaches your kitchen. There are exceptions, of course, but the general picture is grim.

And even if you had a perfect diet of perfect food, your body is so physiologically assaulted by the environment that you'd need supplements just to compensate. The average person is exposed to five hundred chemicals a day. (And we're not even talking, here, about what we knowingly put into our bodies when we eat poorly.)

So in this chapter I (Rob) am going to detail for you the daily supplements I recommend for everyone, as well as supplements aimed at particular symptoms. You'll see some of the supplements discussed in the chapter on the Cleanse again here, treated in more depth, as well as additional suggestions.

START WITH THE RIGHT PRODUCTS

There are many good products out there, but still you must choose your supplements carefully. Steer clear of anything with added sweeteners. Make sure to get products with no alcohol or sugar—two ingredients many companies use as preservatives. Some supplements are contaminated with yeast and fungus, or their spores, and thus are obviously counter-productive at best, and downright harmful at worst. Stay clear of products that contain algae or mushrooms. You should check with any company you buy from to find out if they are sensitive to this issue and what safeguards they have in place. For my money, if they aren't checking on it some-how, that's not a company I'd buy from. Some reliable, high-quality brand names are Source Naturals, Solaray, Innerlight by Darius International, and Nature's Way.

DOSING

You can always take supplements according to the instructions on the packaging. I'll give you my guidelines here. For the rec-ommended supplements that come in capsule form, take one or two capsules at least three times a day. (So, that's up to a maxi-mum of 3000 mg. per ingredient per day, though most you'll be using in combinations so you won't reach the maximum for any one ingredient, and you shouldn't try to.) The liquids you should take under your tongue, three to five drops three times a day. With either form, in serious conditions, you can increase the frequency (keeping the amount the same) to six to nine times a day. Where dosing directions are different, I've noted that in the description.

In the US, the standard capsule size in the supplement industry is 500 mg. They may contain up to 500 mg. of a single ingredient, or a mix of ingredients (some of which may be inert) up to a total of 500 mg. per capsule. Colloid preparations should contain five to ten parts per million (ppm).

I am not actually picky about how much exactly you get of any one ingredient within this framework for capsules and liquids. Most of what I'm recommending comes mixed with other things, and you'll certainly want to take advantage of combinations to keep the total number of capsules you take within reason. The combinations (at least from the good companies) mix ingredients that complement one another, so if you get 25 mg. rather than 50 mg. of one thing, you're probably getting an extra 25 mg. of something else that works similarly in the mix. There is general agreement on what level of each ingredient is effective, and most manufacturers use roughly the same amounts. If you look at all the options on the health food shop shelf, you'll see there's at least a loose consensus about how much you should take. Any place where the specifics matter, I've spelled that out in the descriptions below.

Keep in mind that the doses given here are generalized, and based on a body weight of 11 stone or 70 kg. Everyone's body chemistry is unique, and you may require a greater or lesser amount than what is suggested depending on your weight, body type, health challenges, and sensitivity. If you're lighter, or more sensitive to supplements, stay on the lower end of the range I've provided. If you're heavier, or not seeing the results you want, move to the higher end. You'll find what is right for you. Listen to your body.

As with any supplement program, you should consult your health care practitioner before you begin taking anything.

COLLOIDS

Much of what I recommend is in colloidal, or liquid, almost homeopathic form. Colloids are the smallest biological form of any matter, and are small enough to pass through membranes. In this form, the nutrients bypass the digestive process and are easily absorbed and ready for use by the body. There is no need to waste energy converting them into a usable form, as with standard supplements. This is particularly important if you have an illness, as that will further compromise your digestion, increasing the risk of leaving you malnourished just when you need nutrients most.

Liquids are not necessarily superior to other forms, but they are almost always a good idea. If you use both liquid and capsules, you'll give your body a range of benefits.

Colloidal supplements can be taken any time, away from meals, with doses spaced out over the day. Place drops under the tongue, rather than swallowing them, for rapid, direct absorption into the bloodstream. You can take one supplement right after another this way, though you should leave 15-second intervals between them. Or put them all into a few ounces of pure water and drink your "colloidal cocktail" slowly. Either way you do it, you are aiming for three to five drops three times a day. Keep in mind that less, more often, is better than more, less often. That is, three drops five times a day is better than five drops three times a day.

USING SUPPLEMENTS

When you are shopping for supplements, you may not always be able to find the exact combinations described here. I've generally listed the ingredients in the order of importance, so although I'm a big believer in the power of the combinations, you'll get most of the benefit out of a blend with at least the first couple of ingredients here.

Later in this chapter, we'll get the specific supplements targeted to specific symptoms and conditions. I want to start, however, with the basic things everyone would benefit from taking daily, such as vitamins and, especially, minerals. Nutritional deficiencies can cause their own problems, of course, but on top of that they also increase the toxicity of mycotoxins (and, in a vicious cycle, mycotoxins interfere with nutrient absorption, creating deficiencies). Minerals are especially important because without them, vitamins and enzymes cannot function. Plus, minerals are what is missing from our soil, and thus our food. Exercise flushes them from the body. Furthermore, specific to our purposes here, detoxification (as with the Cleanse, or just shifting to an alkaline diet) requires extra mineral nutrients.

If all that follows seems like too much, remember that you can make quite a lot of progress using just concentrated green powder and pH drops along with a revamped diet. A multivitamin and a multimineral formula compose the next basic layer. Beyond that, the benefits are laid out here to help you decide what is best for you (working along with a health professional). All the supplements here complement the dietary changes you are making, support you while you are making those changes, and protect you when you don't eat right. They are invaluable if you are facing particular symptoms. But you have to determine what works best for you—including how many supplements you can manage.

THE STARS

Here's what everyone should take daily:

Aerobic Oxygen or hydrogen peroxide (H_2O_2), or "pH drops" are safe, stable substances that release oxygen in the body and chelate (combine with) foreign material. Remember, the bad microforms are anaerobic, thriving in the absence of oxygen. The oxygen released by these substances devolves and eliminates yeast and fungus and inhibits spores.

These drops help your body stay alkaline, neutralizing and pH-balancing acidity. You should use a 2 percent solution, which is what almost all products are. Add 16 drops per liter or two pints of pure water—and drink four of them a day (for smaller servings, add four to six drops to eight to twelve fluid oz./250–350 ml. of water). Or follow the directions on the label.

Concentrated green powder. Ideally, at least three of the four liters of water with pH drops you drink should also have a teaspoon of a powdered concentrate of greens and grasses mixed in. Many different companies make this kind of supplement, and many different concentrated green powders are on the market. Single-ingredient powders (just wheat grass or alfalfa, for example) may be easier to find, but you should look for a mixture of grasses and vegetables, which are easily assimilated and rapidly alkalizing, and combinations that are naturally rich in fiber, which bind with and remove toxins from the body. The key ingredients are (organic) wheat grass, barley grass, and kamut grass. They should be mixed with a variety of green vegetables such as broccoli, kale, and spinach—the exact ones are less important than simply getting a wide variety. A formula containing Irish moss, a herb rich in alkaline salts, which can bind mycotoxins and take them out of the body, would be an excellent choice. Sprouted ingredients are good, too, as the sprouting phase is when a plant is at its nutritional peak.

Green powder is nutrient-rich (including easily digested protein) and helps to gently pull your blood and tissues from acidic to alkaline, reaching a natural, ideal pH balance.

A broad-spectrum **multimineral** formula with cell salts. Look for one with a wide variety of minerals and trace minerals, plus the twelve cell salts. You should take the "macro minerals," including calcium, magnesium, manganese, zinc, and iron, plus all of the trace minerals—of which there are about eighty-seven. The wider variety you get, the better, but in case you can't find all eightysomething, be sure to get phosphorus, potassium, zinc, selenium, copper, chromium, and iodine. You should get 1 mg. each of the cell salts (mineral salts, also known as tissue salts, which are at the foundation of every cell, and without which we would die). They are: potassium sulphate, magnesium phosphate, sodium chloride, sodium phosphate, sodium sulphate, calcium phosphate, calcium sulphate, calcium fluoride, ferric phosphate, potassium chloride, potassium phosphate, and silica. Each capsule of the multimineral formula you choose should be 500 mg., as most are.

A broad-spectrum **multivitamin** formula with cell salts. Here again, look for a wide variety of vitamins, ideally combined with cell salts (same as above). At minimum, it should have vitamins A, thiamine (B1), riboflavin (B2), niacin (B3), choline (B4), calcium pantothenate (B5), pyridoxine hydrochloride (B6), biotin (B7), inositol FCC (B8), folic acid (B9), cyanocobalamin (B12), vitamin C, vitamin E, and PABA (vitamin Bx, found in B-complex foods including wheat and rice bran.)

Because the B vitamins are so important, you might consider taking a liquid **B-complex** formula in addition. B vitamins are used in cell formation and are important for cell longevity, and are used in making enzymes. You can use a 500-mg. capsule combining all the B-vitamins (at the usual one to two capsules three times a day), or a liquid, which you can find in colloidal as well as regular form. Whichever kind of liquid you get, the dose is the same: three to five drops three times a day. Or you may want to follow the package directions, particularly for the noncolloidal liquids, as each manufacturer's version is a little bit different.

Essential fatty acids (EFA), found in fish and seed oils, include omega-3 fatty acids and omega-6 fatty acids. Fatty acids are normal parts of human cells, and are especially abundant in brain cells, nerve synapses, visual receptors, adrenal glands, and sex glands. If you are eating lots of healthy oils and have no particular symptoms, you may not need supplements, but the essential fatty acids are so important you should make sure you get plenty, one way or the other.

Let's look at the omega-3 fatty acids first. The omega 3s *eico-sapentaenoic acid* (EPA) and *docosahexanoic acid* (DHA) are antimycotoxic, lowering levels of repair proteins (a sign of mycotoxins in the blood) in the arteries. Repair proteins are also involved in the development of atherosclerosis, so lowering them is a double bonus.

Omega-3 fatty acids offer much more in the way of protection for the heart: They help disperse saturated fatty acids, which like to stick together, thereby avoiding lumps of fat in the blood that dampen electrical charges. They keep blood platelets from getting too sticky, resulting in lower likelihood of clots that can cause heart attacks or strokes. They lower triglycerides by up to 65 percent. They lower cholesterol and low-density lipoprotein (LDL). They lower very-low-density lipoprotein (VLDL) by half. High triglyceride levels, and especially LDL and VLDL, are associated with cardiovascular disease: high blood pressure, atherosclerosis, heart and kidney failure, stroke, and heart attack. Furthermore, our bodies make prostaglandin PG3 from EPA. PG3 helps prevent strokes and heart attacks, lowers high blood pressure (by blocking the production of pressure-raising PG2s found in meat), and prevents pathological blood clotting in the lungs and blood vessels.

And if that's not enough for you, animal studies have shown EPA and DHA to inhibit the growth and metastasis of tumors.

EPA and DHA are found in cold-water fish and other northern marine animals. Trout, salmon, mackerel, sardines, tuna, and eel are the richest sources of omega-3 fatty acids. For those who wish to avoid animal products, there are plant sources for a precursor to EPA and DHA, another omega-3 called alpha-linoleic acid (ALA), including linseed, hemp, walnut, and soybean oils. The body has to work to make EFA conversions like this one, which requires several stages, and in a weakened system it may be better to take the preformed (animal) omega-3s.

Make sure the EFAs you get are fresh to ensure good results. (Break open a capsule—there should be no "fishy" odor.)

Omega-6 fatty acids are found in the seeds of borage, primrose, sunflower, and black currant, as well as in fish oils. Gamma-linolenic acid (GLA) is the primary one. Borage seed oil contains up to 24 percent GLA, evening primrose oil about half that. Borage also contains about 34 percent linoleic acid, another omega-6. Safflower has 79 percent, sunflower 69 percent, almonds 26 percent, pumpkin seeds 42 percent, and rapeseed 28 percent. Hemp oil has three parts omega-3s to one part omega-6s—just as fish oils do.

Omega-3s and 6s have a wide range of health benefits. They can:

- Lower blood pressure, blood cholesterol, and risk of stroke and heart attack.
- Normalize fat metabolism in diabetes and decrease the amount of insulin diabetics require.
- Prevent liver damage due to alcoholism, reduce withdrawal symptoms from quitting an alcohol addiction, and help with a hangover.
- Contribute to building a prostaglandin that helps some schizophrenics.
- Cause weight loss by increasing metabolic rate and fat burnoff.
- Relieve premenstrual breast pain and PMS. Borage oil together with vitamins and minerals has been shown to bring improvement in almost 90 percent of patients.
- Prevent drying and atrophy of tear and salivary glands (Sjögren's syndrome).
- Prevent arthritis in animal studies.
- Improve the condition of hair, nails, and skin.
- Improve some types of eczema.
- Slow or stop the progress of multiple sclerosis, especially

if begun soon after initial diagnosis. (Fish oils have been used with equal effectiveness.)

- Help treat nerve degeneration (diabetic neuropathy) in Type II diabetes, when sugar and saturated fats are removed from the diet as well.
- Kill cancer cells (cells infested with mold and mycotoxins) in tissue culture without harming normal cells.

THE SUPPORTING CAST

If you are ready to expand your use of supplements, the next place to look is at the list of supplements most useful during a Cleanse. They form a core supplementation program.

Noni fruit concentrate, which is antifungal and antiparasitic. It works by activating enzymes and allows the body to renew its cells and rebuild healthy blood and tissue. It also improves digestion and absorption of nutrients, thanks to its enzymatic influence, and helps cells use protein. Noni regulates the health of cellular proteins, as they are used in the creation of different body chemicals. Noni fruit has been used traditionally throughout Polynesia for a wide range of symptoms, including digestive problems, intestinal parasites, skin disorders, allergies, arthritis, and diabetes. The active ingredient, xeronine, is also found in papaya, and is physiologically active in trace amounts. Minuscule amounts occur in practically all healthy cells of plants and animals. Noni fruit also contains significant amounts of xeronine's precursor, proxeronine.

Choose a colloidal supplement. And don't worry, noni is a very bitter—*low-sugar*—fruit. As a result, many supplements are full of added sugar or other sweetener, so be sure to steer clear of those. You also want to avoid any pasteurized noni products. Most likely you'll find it as a powder, in capsules. You may also be able to find a colloidal preparation.

Colloidal silver supports the body's own natural defence system, and is a powerful natural alkalizer. It assists in the organization of cells that make up new tissue. Look for a colloidal preparation.

Chlorophyll. One of the best things you get out of juiced greens is chlorophyll. Concentrated green powder is another excellent source. But you can also get liquid chlorophyll extracts, which you can take separately or use to boost the effects of the juice. (Just don't use it *instead* of juicing.) If you aren't accustomed to the taste of vegetable juice, you may find that a mint-flavored liquid chlorophyll smooths it out quite a bit. (Be sure to avoid the preparations with added sugar, which is especially common when there is also mint added.) Chlorophyll comes as a liquid; add a teaspoon per eight fluid oz. pure water.

Rhodium and **iridium,** which are minerals that come in colloidal form, provide nourishment to cells that have been damaged by mycotoxins, allowing them to recover their ability to communicate with each other effectively.

DNA conductivity was increased ten thousand times when a rhodium atom was added at both ends of the strand. Of course, this was done in a lab, and it has not been determined that the body does this. But it does show that the superconductive (allowing electrical current to flow without resistance) potential of a metal can be biologically active. U.S. Naval Air research has shown that cells in living tissue communicate with each other in a superconductive fashion, but the identity of the superconductors was not determined. While there is still much we need to understand about the process, clearly these conductive metals are effective in stimulating electrical impulses at the cellular level, inducing the flow of electricity in and among cells.

Pine bark extract. One of the most valuable bioflavonoids is pine bark extract, which helps bind up acidity, thereby reducing inflammation (aches and pains) in the body. It has been shown to bind directly with the body's connective tissue, maintaining and repairing it. Pine bark extract is an exception to the rule in that even if it comes in normal-size capsules, most products contain just 25–50 mg. of the stuff, which is fine. You still want to take one to two capsules, three times a day.

An **antimycotoxin formula** combining n-acetyl cysteine, l-taurine, or organic sulphur, all of which are excellent at detoxifying mycotoxins and binding to them to escort them out of the body, as well as expunging acids from the body. You can take them individually or together. I like a combination. Organic sulphur on its own is a popular choice, though if you're going to take only one it should be n-acetyl cysteine. The usual dose applies, one to two capsules three times a day, though if you are using more than one of these individually, you should take less of each, say around one capsule three times a day.

The most important of these three is n-acetyl cysteine, a form of protein, which controls negative microforms and is a powerful antimycotoxin, providing excellent protection against a broad range of toxic hazards (including the toxins acrolein, in barbecue and cigarette smoke and auto exhaust; paraquat, a herbicide; overdoses of paracetamol, a common pain reliever; halothane, an anesthetic; and the side effects of the anticancer drugs doxorubicin and cyclophosphamide). Studies show that n-acetyl cysteine can also bond to toxic heavy metals such as lead, mercury, and cadmium and escort them out of the body.

Glutathione, an important antioxidant with key roles in enzyme activity, is a derivative of n-acetyl cysteine and may be a key to the protective value of the protein. Supplementing with n-acetyl cysteine has been shown to increase glutathione levels in the kidneys, bone marrow, and particularly in the liver, which uses both compounds for protection against mycotoxins.

Research on n-acetyl cysteine focused on acetaldehyde (a primary mycotoxin of yeast and fungus, also found in cigarette smoke) shows just how powerful it is. In a series of trials of many different nutrients, l-cysteine was shown to reduce the deadly effect of acetaldehyde by 29 percent. Glutathione brought it down 33 percent. But n-acetyl cysteine reduced it to zero! This potent antidote to the polluted modern environment is a normal component of the body, but to receive its maximum benefits requires supplementation.

An **antiyeast formula** containing caprylic or undecylenic acid. Caprylic acid controls negative microforms and their toxins. It is an antifungal saturated fatty acid approved by the Food and Drug Administration in 1984 for sale over the counter. Studies have shown patients treated with caprylic acid have completely eliminated fungus from their stool. It can also bring about a remission of symptoms in fungus-related health problems, and appears to be safe and effective with no serious side effects. The most effective caprylic acid formulations are those designed to be released in the colon, where most fungus resides. Along with being effective in eradicating fungus, caprylic acid is also useful after treatment for gout, indigestion, yeast infections, toe fungus, and rashes, as well as for prevention.

Look for 25–50 mg. each of caprylic or undecylenic acid as components of the formula. You'll almost always be using a formula, though you may be able to find one or the other on their own in colloidal form.

CATHERINE AND CHERYL'S STORY

When my daughter Cheryl became so ill that she had to be hospitalized and put on antipsychotic drugs so strong they came with a possibility of causing permanent nerve damage, the doctors admitted they weren't even sure what was wrong. Schizophrenia? Bipolar disorder? A psychotic episode? I knew we were really in deep when one of the many, many doctors I consulted in my search for something better for Cheryl—a psychiatrist so well known that if I used his name you might recognize it— told me, "They'll give her condition a name and they'll give her a drug, but they don't really know what they're doing or what they are talking about." This from a man whose business it is to give these things a name and pre- scribe a drug, who in fact has gotten famous for doing so! He suggested I find a residential facility and check her in for at least a year—in part to get some rest for myself!

I wasn't about to send Cheryl off into the hands of still more doctors and psychologists, none of whom seemed to have a clue what was really going on or what to do about it that might actually be productive. I turned my search to "alternative" medicine, and found a psychiatrist who deals with depression and "mental ill- ness" through nutrition. Through my own research, I had already started Cheryl on an all-organic, high-pro- tein diet and approximately two hundred supplements a day. This new doctor agreed with that approach, and added weekly B12 shots. Cheryl seemed to be getting better.

Eight months and six thousand dollars into this regime, Cheryl was still having violent mood swings and depression, and she finally slipped back into psychosis. She ran away and it was a year before we could get her back into the hospital again. I had to smuggle supplements in to her in plastic containers—as "malteds," ironically. No supplements were allowed for the "mentally ill" patients—only dangerous antipsychotics, and as much sugar and caffeine as they liked. And don't even get me started on the food. I could never approach the subject of nutrition with her doctors, never mind tell them she was taking hundreds of vitamins a day. When Cheryl was released from the hospital faster than many other patients, many of whom were less ill than she was at the outset, the doctors attributed her rapid recovery to their new wonder drug.

After she was released, I took her to Mexico for live cell therapy to the tune of fifteen thousand dollars. It allowed her to get off the drugs, but it didn't stop her mood swings or depression. She cried almost every day from May to September. So did I.

Throughout all this, my own health seriously degenerated. I had been neglecting my body, and the incredible stress took its toll. I was swollen all over (from yeast, I now know). I had pain all over my body. My vision was often so bad I could hardly see. I had no energy to speak of.

Finally, I ran across Dr. Young. Both Cheryl and I

had live blood analysis and learned about the parasites in her blood that cause anxiety and depression. We learned about this program, and I knew that, at last, we had found the solution.

We both did the Cleanse and started taking several antifungal/antimycotoxin formulas, including colloidal caprylic acid, undecylenic acid, germanium, and n-acetyl cysteine, as well as rhodium and iridium for the brain, and the antiyeast enzymes amylase, bromelain, papain, and lipase, multivitamin with cell salts, l-taurine, and omega-3 oils, then continued to the full diet.

Cheryl's depression stopped. Her mind is clear. She laughs regularly.

I am ecstatic. My "baby" is well. My health is better than it's been in years. My mind is clear and keeps getting clearer. I am more perceptive, visually and mentally. My skin feels wonderful and I can see my cheekbones again for the first time in twenty years. I have greatly increased energy, and a general sense of well-being.

Our kitchen is filled with bowls of sprouts of every variety. We are eating almost totally raw, and she is concocting many wonderful and some, shall we say, "interesting," recipes. We hope to be starting an "Un"Cooking class soon. Cheryl has a vision of opening a "Diet Center" based on the principles of The pH Miracle, where people can come and get well. Thanks to this program, I know whatever her dreams, she'll be able to make them come true.

Undecylenic acid, made through vacuum distillation of castor bean oil, is another fatty acid proven to eradicate or devolve yeast and fungus and counter their toxins. Listed in the *U.S. Pharmacopoeia* for use as a topical antifungal, undecylenic acid can also be used orally for treatment of psoriasis, neurodermatitis, and intestinal fungus. Some studies have shown that undecylenic acid is even more effective than caprylic acid.

ENZYMES

Bromelain is useful both for its digestive enzymes and for its antiyeast properties. You'll probably be able to find it in combination with caprylic and undecylenic acids, which is an excellent way to take it. Bromelain, an enzyme from fresh pineapple and papaya, is a safe and highly effective antifungal, antibacterial, and antiparasitic. I recommend bromelain primarily because it helps break down encrusted waste material and dried mucus, which is particularly helpful in the large intestine, where most fungus is found. Bromelain removes dead bacteria and fungus and boosts the healthy flora reintroduced by probiotic supplements. As these many supplements do their work, devolving and eliminating yeast and fungus, bromelain loosens and dissolves remaining cell membranes and other debris.

Bromelain is also an anti-inflammatory that can reduce swelling, pain, and inflammation of injuries, ulcers, and joints. It works by balancing hormone-like substances in the body called prostaglandins, which in turn help control inflammation. Bromelain improves cardiovascular health by cleaning the plaque out of the blood vessels and reducing the tendency for blood cells to stick together—reducing the risk of stroke and heart attack. Bromelain also enhances digestion, particularly of protein.

You can take up to 500 mg. capsules of bromelain—in the usual one to two capsules three times a day dosing—though it is not often found that way and isn't necessary anyway. Bromelain is more often found in combination with other digestive enzymes (see below). Look for a formula with at least 50 mg. of bromelain in it.

Digestive enzymes, including amylase, protease, lipase, papain, as well as bromelain, are crucial if you are still transitioning, especially if you are eating any animal proteins. Once your diet is entirely alkaline, and full of high-water-content, low-sugar foods, you are unlikely to need additional digestive enzymes. But if you become nauseated when eating, have several food allergies, have heartburn, have wind, bloating, belching, abdominal pain, or cramps after eating, or are often constipated, these enzymes may help you restore good digestion. And even if your diet is ideal, it can be wise to take enzymes when you are eating cooked food.

One word of caution here. These enzymes are frequently prepared on beds of yeast. It is very important that purification and testing are done during manufacture to ensure that no organisms, spores, or other by-products remain in the final product. Check with manufacturers to make sure they have safeguards in place.

Herbal enzymes bring much the same benefit. Look for a formula including papaya leaves, peppermint leaves, ginger root, catnip, and fennel; you might find these in combination with bromelain as well. Papaya is particularly useful in that it helps to completely digest proteins, breaking them down into their amino acid building blocks, making them immediately usable for the body.

Antimycotoxic enzymes include superoxide dismutase (SOD), catalase, glutathione, glutathione-S-transferase, glutathione peroxidase, and methionine reductase. These are not digestive enzymes; rather, they chelate (bind to) acids, allowing them to be excreted from the body. A concentrated sprouted organic wheat (wheat grass) supplement is an excellent source. You can also take capsules, in the usual way. Look for a combination with 25–50 mg. of each enzyme, for a total of 500 mg. per capsule (may be difficult to find in the UK).

These enzymes are also antioxidants, or neutralizers of free radicals—highly reactive, potentially damaging substances created as a by-product of oxygen use. Free radicals are necessary for some chemical reactions, and are even used to counter some other harmful elements. However, the amount of free radicals can easily get out of control, and then they can damage healthy cells and essential substances. It only takes one free radical molecule, for example, to rupture a cell membrane. Such damage weakens us, accelerates aging, and puts greater demands on repair processes, enzyme production, the immune system, and energy reserves. Antioxidant nutrients essentially "mop up" excess free radicals and so prevent damage.

I'm betting you're pretty familiar with all of that by now, with all the recent hype about antioxidants. And there's no doubt they're beneficial, and free radicals are potentially dangerous. But all of that is missing an important point: The real culprits in most of the damage attributed to free radicals are fermentation processes, and the mycotoxins they produce. Morbid fermentation results in substances being oxidized (creating free radicals)—but it is in response to the mycotoxin by-products of the fermentation that the body actually creates the free radicals.

Furthermore, if the immune system is going to overreact, it is likely to happen in a crisis, under tremendous stress. Without the stress of mycotoxins on the body, stress overload would occur much less easily and often. The immune system is already busy enough in a healthy body. Don't push it! In any case, as I've said, the immune system cannot produce wellness.

So provide the body with plenty of enzymes, yes—but do it for their antimycotoxic properties as much as their antioxidant benefits.

THE CAMEOS

The following nutrients are helpful for bringing your body back into balance. Once you are there, they are optional, though you'll do well to continue with them.

All these would be great on their own, but they are often found in combination formulas as well. When you are evaluating products and various combinations, these are some of the most beneficial ingredients to look for.

Olive leaf extract acts against morbid microforms, is an excellent antimycotoxin, and has antibiotic properties. It also benefits the cardiovascular system by protecting HDL cholesterol ("good cholesterol") from oxidation. Botanists believe it is the chemical compound oleuropein's presence throughout the olive tree—in the wood, fruit, leaves, roots, and bark—that protects it from insects and bacteria. Furthermore, calcium elenolate, made from one of oleuropein's breakdown products (elenolic acid) is a major destroyer, or growth inhibitor, of many kinds of microforms. Another by-product, aglycon, has a similar inhibitory effect. Olive leaf extract acts against various fungi, as well as salmonella and *Staphylococcus aureus* bacteria. You can use olive leaf extract on its own, though you should note that this is another exception: Solo capsules will contain just 25–50 mg. of olive leaf extract, and that is just fine. You should look for similar amounts in each 500 mg. capsule of a combination.

Garlic has been widely used in health and medicine for centuries. For example, both the Roman poet Virgil and the Greek physician Hippocrates mention it as a remedy for pneumonia and snakebite. Though it is silent on the snakebite issue, modern science does tell us garlic is a good antifungal and antibacterial agent and inhibits yeast and mold as well as fungus and bacteria. It has been shown to be effective against the bacteria *Staphylococcus aureus* and *E. coli,* in particular, as well as on *Candida albicans.* Of these three common organisms, *Candida* has been shown to be the most sensitive to garlic juice.

Even small amounts of garlic are effective, but the chemical component that is most therapeutic (allicin) also gives garlic its strong odor. So beware of the "odorless" formulations (though they may still provide benefits to cholesterol levels and fat metabolism).

Garlic supplements are particularly important if you don't like the taste or smell of garlic, and so aren't getting it in your regular diet.

Butyric acid, another short-chain saturated fatty acid, helps chelate (bind up) mycotoxins that increase low-density lipoproteins (LDL cholesterol). It is healing to the mucous membranes of the stomach and small and large intestines. It can, for example, repair the damage *Candida albicans* creates boring holes in the intestinal walls. Butyric acid also boosts immune function by detoxifying the lymph system of yeast and fungus and their associated mycotoxins. Butyric acid, which comes as a liquid, alone, or with other antifungals, can be hard to find. As long as you are getting caprylic or unde-cylenic acid, you don't have to worry about getting butyric acid if you can't find it.

Thioctic or **lipoic acid** chelates mycotoxins collecting in the liver and normalizes liver enzymes. It has also been shown to remove mercury, arsenobenzoles, carbon tetrachloride, and ani-line dyes. Experiments demonstrate a vast increase in oxygen supply and use with lipoic acid treatment. This liquid can be another tough one to find. Your best bet may be as a combina-tion with caprylic or undecylenic acid, and again, if you are getting one or both of those two, this one is not crucial.

Organic **germanium** is a metallic element that helps elim-inate yeast and fungus, thanks to its promotion of increased production of interferon, which has antitoxic and antiparasitic activity. It also enhances metabolic chemical reactions based on oxygen (which, in the human body, is most of them). Organic germanium also stimulates electrical impulses on a cellular level, helping the body discharge certain unwanted electrical fields and allowing much-needed current to flow through. In other words, it helps establish the desired electri-cal balance. That's crucial because electricity provides fundamental organization and control in the body—like a framework for all other processes. Germanium capsules gen-erally contain just 25–50 mg.; if you get a combination look for that much germanium.

Rare metals (in addition to rhodium and iridium, above), including gold, gallium, osmium, ruthenium, palladium, and platinum. These are all colloids, so you should use three to five drops three times a day, or follow package directions.

Gold supports the body's own natural defence system. It has been used successfully to treat arthritis, skin ulcers, burns, certain nerve-end operations, various types of punctures, obesity, and inoperable cancer, but fell into disuse with the advent of antibiotics and other (toxic) drugs. Gold is a conductor of electricity, which may help in cellular communication, metabolism, and regeneration. Research has shown gold has the potential to repair damaged DNA. Gold can have a psychologically balancing and harmonizing effect, easing depression, seasonal affective disorder, melancholy, sorrow, fear, despair, anguish, frustration, and even suicidal tendencies. The preparation of gold to look for is expensive jewellery. Just kidding! Seriously, the colloidal form is what you need.

Gallium helps form antitumor compounds. It has specific areas of enzymatic activity in the human brain and has been reported to reduce the rate of brain cancer in laboratory animals. British research shows that pregnant women taking supplemental gallium reduced the rate of brain cancer in their children.

Osmium, ruthenium, palladium, and **platinum**, as well as rhodium and iridium, have extraordinary electrical conduction properties. Like germanium, they are electrostimulators, but with a different speciality: Increasing the ability of DNA to conduct electricity and enhancing communication among cells. They also stimulate metabolism. Choose a colloidal preparation.

Research at the Bristol-Myers-Squibb labs indicates that using precious metals in the presence of cancer can correct altered DNA. Scientists coupled these elements with the cells via a transfer of light—encoded bursts of ultraviolet laser light. Electrons that flow through a superconductor pair off and convert into light. Superconductors assist light transfer, and gallium, gold, rhodium, iridium, osmium, ruthenium, palladium, and platinum increase the light found in the human body.

EDNA'S STORY

All I wanted to do was sleep. I would sit in my office and find I could not stay awake. I'd go home at night and fall asleep on the floor until it was time to go to bed. I was not functioning well or doing the work I needed to do. I couldn't even attempt to exercise. I was stressed and pale, enough so that people would ask me if I was okay.

I was ready to try anything, so when live blood analysis was suggested to me I thought that, eccentric as it sounded, I'd just go and see what it was about. Knowing what I know now about what I was (and wasn't) eating, I shouldn't have been surprised that the analysis showed acute yeast/fungus imbalance, adrenal stress, and irregularly shaped red blood cells.

I went on a Cleanse with supplements and juiced green vegetables. After three days, I added all sorts of vegetables, and turkey and fish. I ate stir fry for dinner, nibbled on carrot and celery sticks for lunch, and had vegetable juice for breakfast.

I started taking vitamin B5, beta carotene, a multimineral, pH drops, undecylenic and caprylic acid, bromelain, citron, chromium, vanadium, omega-3, borage and fish liver oils, colloidal silver, and a few herbal combinations with at least another twenty ingredients. I couldn't believe how many pills I had to take! By the time I took my supplements and a little juice, I would be full.

I stayed on this regimen for two months with no cheating. Gradually, I felt better and better. After the first week, the color returned to my face and I had more energy. I slept better at night. The most important thing to me was that I could go home and prepare dinner for my family of nine. Along with feeling better and having more energy, there was another benefit: I lost sixty pounds. My husband says that even my eyes are brighter.

I went shopping and bought a dress with a straight skirt. I haven't worn a straight skirt in twenty years. My closet was crammed with clothes from size eight to eighteen, and I just cleaned out everything above my current size: Nine!

At my follow-up blood test, I was pleased to actually see my improvement on the monitor screen. This time I had perfectly round red blood cells, and the fluid was free of most bacteria and had very little yeast.

I really do enjoy what I eat now. Before I would never eat raw nuts or avocado. They tasted awful to me. Now my tastebuds have changed, and even vegetables taste sweet! Tomatoes are a real treat for me.

This program brought back my sense of well-being and allowed me to lose weight I haven't been able to lose any other way. It literally saved my health.

OPTIONAL

These supplements are not for everyone, but if they suit your situation, you'll find them to be invaluable:

Probiotics can be useful in re-establishing healthy micro-forms in healthy size populations in your digestive tract. Not everyone needs them. If you have been constipated, under-weight, overweight, or on antibiotics in the last year, or you have poor digestion, however, give them a try. Mix one to two capsules in an eight fluid oz./250 ml. glass of pure water and let it sit for eight hours to increase the microflora considerably. Drink half an hour before each "meal" during a Cleanse, or real meals thereafter. After that, if your digestion and elimination are good, you can stop the formula. Otherwise, continue as needed.

You should use a product that combines different bacteria—the ones you'll see most commonly are **acidophilus, lactobacillus,** and **bifidus**. You should get 50–250 mg. of each, depending on what else is in the mix.

Another option is to do a probiotic formula enema each morning (starting during a Cleanse, if you want to). Mix four capsules of probiotic formula with 12 fluid oz./350 ml. of pure water and let sit for eight hours, or overnight. Use half the mixture and hold the enema for 15 minutes. Repeat after half an hour.

An **antiparasite** formula containing **black walnut hulls.** Take four capsules three times a day for 10 days, then break for four days. Repeat this cycle at least six times. Although you can get black walnut hulls alone, I prefer combinations that include it.

ADDRESSING SYMPTOMS WITH SUPPLEMENTS

Beyond the basics, the correct supplements are powerful tools in clearing up specific symptoms. You can add the supplements below to the routine you've already established for yourself, thereby tailoring it to your specific situation. Use them until you leave the symptoms behind, taking them occasionally as necessary for prevention thereafter. If you need the same supplement for more than one condition, don't multiply the dose—but do use the highest dose given.

Sinus symptoms: Dissolve two capsules of an antiyeast formula in a bottle of saline solution (available at pharmacies). Spray into nostrils at least three times a day, following package instructions. You can also use one or two drops of colloidal silver or liquid pine bark extract in each nostril once or twice a day. If you can't get liquid pine bark extract, use the powder mixed with pure water, put in each nostril once or twice a day. Sinus symptoms should steadily disappear on their own as toxins are cleared from the body and good digestion is restored.

Eye or ear problems, including cataracts, glaucoma, redness, blurred vision, poor eyesight, ringing in the ears, earaches, soreness or swelling of the ears, eardrum damage, hardness of hearing, and (in rare cases) loss of hearing: Use one drop of colloidal silver topically (directly in the eye or ear) three times a day.

Congestion: You have three options to experiment with, to see which works best for you:

1. If you aren't already using it, add three to five drops of Aerobic Oxygen in eight fluid oz./250 ml. pure water three times a day.
2. Use three to five drops of colloidal silver or liquid pine bark extract three times a day, under the tongue.
3. Use seed oils, such as linseed oil and borage oil, in capsules as you would any supplement. Or take one teaspoonful three times a day. Or use more in your diet!

Liver stress. The liver, which filters toxins from your body, should itself be detoxified at least three or four times a year. Look for a liver formula combining glandular liver, dandelion root, red clover, chapparal, yellow dock root, cascara sagrada bark, licorice root, sarsaparilla root, celery seed, burdock root, echinacea, Oregon grape root, stillingia, prickly ash bark, buckthorn bark, cayenne, kelp, and wild yam root. Take two capsules at least half an hour before each meal (three meals a day), with an eight fluid oz./250 ml. glass of pure water, for a total of 180 capsules in 30 days. If you are using a probiotic formula, take the liver formula with that. This approach can also help in cases of hepatitis, cirrhosis, and jaundice.

(Glandulars [such as glandular liver, here, glandular lung, below, and others in this section] are in fact animal products, usually from cows. They are in the formulas specifically as messengers of sorts: They carry nutrients to specific areas. That is, they bring nutrients to the part of the body the gland came from. For example, if you tagged the formula above with radioactive isotopes and then followed it through the body, you'd see it go directly to the liver. [And such studies have been done.] You can get similar products minus the glandulars if you want to be scrupulous about avoiding animal products, though they won't be as targeted.)

Lung problems, including pneumonia, asthma, bronchitis, croup, tuberculosis, colds, flu, hayfever, and emphysema: Look for a lung formula including glandular lung, pleurisy root, wild cherry bark, slippery elm bark, plantain, mullein leaves, chickweed, horehound, licorice root, kelp, cayenne, and saw palmetto. Take two capsules with each meal. You can also use five to ten drops of colloidal silver or pine bark extract in a respirator. You can buy a respirator at your local pharmacy. Add water and colloidal drops and it creates a mist in the air for you to breathe in, which is great for lung or nasal congestion.

SHIRLEY'S STORY

I finally went to the hospital just after Thanksgiving. I just couldn't seem to get rid of what I thought was a terrible cold, and had been coughing almost continuously day and night for about a week. So the doctors admitted me for tests. At first they thought it was pneumonia. Then the lung specialist thought it was another kind of infection in my lung. The medication he prescribed seemed to help and I went home after a few days.

Just before Christmas my doctor called to ask me to return for more tests after the holidays. Back in the hospital, the doctors found two "spots" on my liver, and a very low sodium count, and they told me they suspected I had cancer! The next day they found a mass on my lung. Then a CAT scan found a tumor on the right side of my brain and another behind my left eye. They finally traced the origin of the cancer to my bronchial tube.

The doctors told my daughters—but not me—that I had small cell cancer, a very aggressive form. If I took chemotherapy, I'd have two to three years to live. Without it, I'd be looking at six months. The doctors asked my daughters not to give me the particulars about how long I had because they didn't want me to lose hope.

I had my first chemotherapy treatment, which took three days. I then started radiation treatments for the tumors in my head—fifteen treatments over three weeks. In the middle of all this, my daughter came to stay, loaded down with bottles of pills and books and tapes about Dr. Young's principles. She totally changed my diet and started me on a bunch of supplements. She kept about ten little brown bottles of what she called colloids on my dressing table and gave them to me before and with meals, as Dr. Young recommended, until I learned how to take them myself. I began to notice a different sensation in my head as well as in my chest.

After about two weeks of my daughter's pH Miracle treatment, I had my first appointment with my doctor since leaving the hospital. He told me my blood work came back normal. Seeming a little confused, he asked if I had had radiation to my chest. I hadn't. He told me that the mass in my bronchial tube was gone—and that if the cancer was not in the place of origin, it was probably nowhere else to be found in my body!

For the second time talking with my doctors (after my diagnosis), I couldn't believe what I had just heard. It took a few hours for it to really sink in. What a weight had been lifted off my shoulders—and my family's! How we celebrated!

Still, I took two more chemotherapy sessions (I guess my doctors couldn't quite believe it either). My blood work continued to come back close to normal, which is highly unusual. At my next doctor's appointment, a month after the last check-up, I got astounding results from my CAT scan: The tumors in my head were definitely gone! My doctor discontinued my chemotherapy and told me to check in with him in another two months. His last words are still ringing in my ears: "Whatever you're doing, keep on doing it!"

Pancreas problems, including hypoglycemia and hyperglycemia (diabetes): Look for a pancreas formula including glandular pancreas, uva-ursi, dandelion root, parsley, gentian root, huckleberry leaves, raspberry leaves, buchu leaves, saw palmetto berries, kelp, and bladderwrack. Take two capsules with each meal, and three to five drops of liquid or colloidal chromium and liquid vanadium before each meal, under the tongue. Vanadium helps to uptake the chromium to help in making effective glucose and insulin interactions.

Adrenal stress (symptoms include insomnia, fatigue, low blood pressure, poor circulation, feeling cold all the time, getting lightheaded when standing, arthritis in joints and back, drowsiness or sleepiness in the afternoon, chronic pain): You have four options, which you can take together:

1. With each meal, take two capsules of an adrenal formula including glandular adrenal and pantothenic acid (vitamin B5).
2. Take three to five drops under the tongue, three times a day, of an antimycotoxin formula containing n-acetyl cysteine, l-taurine, and organic sulphur. Caprylic and undecylenic acid, pine bark extract, and grape seed extract would be good additions.
3. Take three to five drops under the tongue three times a day of a formula containing chromium and vanadium.
4. Take three to five drops of a liquid B-complex three times a day, under the tongue.

Thyroid problems: Look for a thyroid formula including kelp (to deliver iodine), gentian root, saw palmetto berries, cayenne, and Irish moss. Take one to two capsules with each meal.

Digestive aids: To relieve intestinal wind (flatulence), enteritis, colic, and heartburn, look for a formula including papaya leaves, peppermint leaves, ginger root, catnip, fennel seed, and saw palmetto berries. Take two capsules with every meal. You then have two other choices:

1. Take one to two capsules noni fruit concentrate before meals with a small amount of water and one to two capsules of an antiyeast formula containing undecylenic and caprylic acids and enzymes such as bromelain and papsin, and herbs that aid in digestion half an hour after each meal.
2. For digestion of fats, with each meal, one to two capsules of noni fruit concentrate and an antimycotoxin formula containing n-acetyl cysteine, l-taurine, and organic sulphur, at least, and perhaps caprylic and undecylenic acids, pine bark extract, and grape seed extract.

Lymphatic blockage: You have two choices:

1. Take one capsule in the morning and another in the evening of a lymphatic formula including fish liver oil, beta carotene, dandelion root, eyebright, marshmallow root, licorice root, and parsley.
2. Although this is not a supplement, it is a very helpful approach to lymphatic blockage, so I wanted to include it here: A series of at least 24 lymphatic massages—massages that move the lymphatic fluids and help to move toxins out of the lymph nodes by stroking with the hands in the direction of the lymphatic vessels—no downward movements. Lymphatic massages proceed from the feet to the legs to the torso toward the heart, from the lower back to the upper back over the shoulders toward the

heart, and from the fingers to the arms to the shoulders to the heart. Consult a massage therapist experienced in this area. Regular massage therapy can also be very beneficial. Even easier, perhaps, is daily dry brushing of the skin. Simply get a skin brush at a natural food shop or body shop and brush away, wet or dry—but always toward the heart.

Kim's Story

I'd been under extreme stress for a long time, and it started to take a physical toll, including high blood pressure and erratic heartbeats and fibrillations. I've also struggled for years with a very painful left breast and lymph node in my left armpit, and very low energy that rendered me dysfunctional to a large extent.

The pain became progressively worse, and I discovered a lump in my left breast. Wisely or unwisely, I chose not to have a mammogram, opting to honor my intuition and beliefs instead. I wasn't prepared to have a biopsy, needle biopsy, drugs, surgery, chemotherapy, or anything that might further endanger my life, so the mammogram wouldn't have served a useful purpose anyway. I realized I was taking a risk, of course, but I wanted to give my body the best chance of being healed in a natural and holistic way.

I started taking colloidal silver after a friend told me he had learned from Dr. Young about its being a broad-spectrum natural antibiotic that fights yeast, fungus, parasites, and viruses. I also learned about the herbal tea Essiac—according to the article I read,

President Kennedy's personal physician had cured himself of cancer by only taking Essiac. I took the colloidal form rather than having to use the tea. Then from Dr. Young I learned about antiyeast enzymes, iodine (for my hypothyroidism), the antiyeast/fungus combination of caprylic and undecylenic acids, and pine bark extract.

To my amazement and delight, four days after starting to take the Essiac and caprylic acid, my lump had totally disappeared! My energy level is definitely improving. I'm still struggling with some health problems, but I realize this is a process, and as I embark on an alkaline diet like The pH Miracle, and get more rest, I know my health will get better and better.

Weight control and fat and glucose metabolism: You have several options—you can mix and match:

1. Take one to two capsules of a multivitamin and one to two capsules of a mineral formula daily. Getting your nutrients in a capsule that is predigested saves energy, and there will be little or no acid produced in getting the nutrients.

2. Take two to three capsules of pine bark extract one hour before each meal and three to five drops of colloidal chromium and vanadium three times a day, under the tongue (use capsules if you can't find it in colloidal form).

3. Take one to two capsules, three times a day, with meals, of a pituitary/thyroid formula containing kelp, gentian root, saw palmetto berries, cayenne, and Irish moss.

4. For women: Take a liquid amino acids formula (look for, among others, lysine, methionine, arginine, leucine, tyrosine,

tryptophan, and phenylalanine), colloidal calcium, and colloidal boron, three to five drops each, three times a day, under the tongue, and one to two capsules of a women's formula including glandular ovary, glandular uterus, black cohosh root, licorice root, raspberry leaves, passionflower, chamomile, fenugreek, black haw bark, saw palmetto berries, squaw vine, wild yam root, and kelp.

For men: Take a liquid amino acids formula (look for, among others, lysine, methionine, arginine, leucine, tyrosine, tryptophan, and phenylalanine), colloidal zinc, and colloidal vitamin B6, three to five drops each, three times a day, under the tongue.

5. A formula containing the fat lecithin.

6. Citron or garcinia cambogia, an alkaline bitter fruit, helps in reducing acidity and thus the need for the body to retain fat.

7. Here's another one that isn't a supplement: Follow the diet portion of this program strictly! Your weight will naturally control itself once you are eating alkaline. You will also find drinking 8 fluid oz./250 ml. of dark green vegetable juice six to eight times a day, as in the first part of the Cleanse, helpful.

Joint and muscle pain: You have four options:

1. Take three capsules of a calcium and four capsules of marine lipids/borage oil formula with each meal. To make sure it combines properly, take the oil with meals made up of vegetables or vegetable juice.

2. Take three to five drops of colloidal calcium, colloidal boron, and an antimycotoxin formula (with n-acetyl cysteine, l-taurine, and organic sulphur, and perhaps caprylic and

undecylenic acids, pine bark extract, and grape seed extract), three times a day, under the tongue.

3. Take a joint and muscle formula—including colloidal glutathione (three to five drops, three times a day under the tongue), calcium and magnesium (two to three capsules three times a day with meals), and zinc (two to three capsules three times a day, with meals). For acute or chronic conditions, use four to six capsules three times a day.

4. Take a formula containing the herb yucca, which reduces inflammation, soreness, and swelling.

Reproductive organ disorders: You have two choices:

1. Take three to five drops colloidal rhodium and iridium three times a day, under the tongue.

2. Women: Look for a "female tonic" including glandular ovary, glandular uterus, black cohosh, licorice root, raspberry leaves, passionflower, chamomile, fenugreek, black haw bark, saw palmetto berries, squaw vine, wild yam root, and kelp. Take two to three capsules with each meal, three to five drops liquid pregnenolone, colloidal calcium, and colloidal boron three times a day under the tongue. A soy sprouts formula (made by dehydrating say, 20 oz./450g. of soy sprouts into 1 oz./30 g. of the supplement via low-heat dehydration) can be substituted for pregnenolone by taking three to five drops three times a day with meals, as can progesterone cream used as directed.

   Men: Take one to three capsules of a men's formula including glandular prostate, parsley, saw palmetto berries, corn silk, buchu leaves, cayenne, kelp, and pumpkin seeds with each meal, and three to five drops each of colloidal zinc, colloidal vitamin B6, and a liquid amino acids formula (look for, among others, lysine, methionine, arginine, leucine, tyrosine, tryptophan, and phenylalanine), three times a day, under the tongue.

Infectious and degenerative symptoms, such as AIDS and cancer: Try colloidal formulas of osmium, ruthenium, and palladium; co-enzyme Q_1 (CoQ_1, also known by the acronym for its hugely long chemical name, NADH); the antiyeast undecylenic and caprylic acids; and enzymes such as bromelain and papain.

Toxic stress (oxidation and mycotoxins): Try colloidal or liquid vitamin C, echinacea, glutathione, calcium, boron, silver, lithium, selenium, pregnenolone, pine bark extract, progesterone (for women—who then don't need the pregnenolone), fish liver oil, beta carotene, dandelion root, eyebright, marshmallow root, licorice root, parsley, n-acetyl cysteine, l-taurine, sulphur, caprylic and undecylenic acids, grape seed extract, a multimineral, and a soy sprouts formula. Progesterone comes over the counter in a cream. Soy sprouts are another natural source of hormones.

Supplements are powerful allies in this program. They will maximize the results you get from the Cleanse and an alkaline diet. But there is no such thing as a magic bullet—no one thing is going to solve all your problems for you, or keep you healthy forever. And as powerful as they are, they are no match for that great scourge—the typical Western diet. If you don't change the way you feed your body, any supplements you take will be overworked. The combination—diet *and* supplements—is the key.

Chapter 11

Putting It Together

The more you follow this program, the easier it becomes to follow. Your body adjusts to it, and so do you. It will come to seem second nature. But right now, you've got a steep learning curve. We're throwing a lot of information at you at once, and you are probably looking at pretty significant changes in the way you live your life. Experience is the best teacher. But while you're gathering your experience, we want to give you some practical strategies for implementing this program. You'll learn here how to stock (restock, really) your kitchen, with food and equipment, as well as how to grow your own sprouts, and how to dehydrate foods. Finally, Shelley will share some of her "secrets" for using the program in your ordinary, everyday, no-doubt-already-busy life.

STOCKING THE PANTRY

You're preparing to embark on the adventure of a lifetime—the adventure of a health-generating lifestyle. You have heard the science, and now you want to know how to continually live that science from the kitchen. When you take a look at what is in your house right now to eat, however, you're likely to face the grim reality that much of what you have isn't good for you after all, now that you understand the New Biology. You probably have a lot that is flat-out lacking in nutrition and fiber, and more that has a very acidic effect in the body.

To make it simple to follow this program, chances are you're going to need a makeover. But once you've transformed your kitchen, you'll find it easy, simple, and natural to follow the principles of a healthy alkaline diet. One key is to keep all the basic items you'll use a lot of on hand in your pantry and fridge, so you can prepare a delicious—alkaline—dish at the drop of a hat.

If you poked around my kitchen for a few minutes, here's what you'd find:

Spices. Just in case I can't get fresh ones. I keep dried spices, which I buy in bulk. I am always stocked up with garlic powder, onion flakes, cumin, basil, coriander, cayenne pepper, turmeric, cinnamon, curry, and parsley. I also always have an alkalizing salt. I prefer Real Salt™ brand but alternatives include vegetized salt (which has dehydrated veggies in with the salt), and sea salt with organic veggies (check your health food supplier for salts with different seasoning combinations). Another good choice includes Bragg™ Liquid Aminos as a substitute for soy sauce (it is still slightly acidic, but is far better than fermented soy sauce). Beyond that, I like to use premixed spice combinations to take the guesswork out of things, avoid unnecessary clutter, and keep it simple! I'm a fan of the Spice Hunter brand (unavailable in the UK—but look for organic herb and spice combinations such as Italian seasonings, Herbes de Provence, curry or Cajun). You can make the same dish seem really different depending on what you shake over it.

Seeds. I keep a supply of (raw) linseed, sesame, sunflower, alfalfa, and pumpkin seeds, as well as sprouting combinations. I also keep raw tahini (sesame seed paste) around.

Nuts. I always have raw nuts on hand, too, usually almonds, hazelnuts, pine nuts, brazil nuts, and macadamias, as well as fresh, seasonal pecans and walnuts and raw almond butter.

Grains. I keep spelt, buckwheat, millet, kamut, quinoa, brown basmati rice, and amaranth handy, as well as unsalted brown rice cakes, quinoa and spelt pasta, buckwheat, rice and soba noodles, sprouted wheat tortillas, and flours (whole wheat, unbleached white, spelt, brown rice, rye, and millet in particular—soy and rice flour would be okay, too, as long as they are not stored for a long time). I also keep unleavened whole wheat bread in the freezer (from the freezer section of the health food shop).

Beans. I usually have soybeans (sometimes known as edamame beans—look for them in the frozen foods section of your shop), adzuki beans, lentils, mung beans, cranberry beans, black beans, black-eyed peas, chickpeas, pinto beans, and kidney beans. I almost always have hummus in my fridge, too.

Sea vegetables. I often use nori sheets, dulse flakes, and aramie, so I keep them around.

Tortillas. Look for sprouted wheat varieties. Some wraps restaurants may sell you their tortillas, or you can make your own (see recipe on page 305).

Healthy oils. I always have a variety available, especially virgin cold-pressed olive oil, grape seed oil, sesame oil, and linseed oil, and a blend (like Udo's Choice® or Essential Balance, under the brand name Omega Nutrition, or Arrowhead Mills). Hemp oil has quite a strong flavor, but I use it on occasion, too. Look for Essential Balance in the refrigerated section of your health food store, and keep it in your refrigerator or freezer. It is a wonderful flavored oil combining organic linseed, sunflower/safflower, pumpkin, borage, and sesame oils. It's the only oil organically processed, made from all organic seeds, and completely unrefined. It comes in a black bottle to keep light from damaging it.

"Milk." For when I need it and don't have time to make my own, I keep almond, soy, or rice milks around.

Water. I keep distilled or reverse-osmosis/purified water in gallon containers.

Soy. I always have some tofu handy, including baked tofu, and I often have soy burgers in the freezer.

Produce. Since produce must be fresh, I obviously don't store it indefinitely—although the stuff that keeps, such as garlic, ginger (I keep a whole root in the freezer and grate it as necessary), onions, fresh and dried chili peppers, lemons, and limes, I buy enough of to always have around. And since I've always got something sprouting, I have a fresh harvest on any given day. (My favorite is a high-protein mix of mung beans, adzuki beans, lentils, peas, and sunflower seeds, and I often use a mix of alfalfa and other small seeds, too.) In my pantry, I'm never without sun-dried tomatoes packed in olive oil, roasted bell peppers packed in olive oil, and vegetable broth (make sure the one you choose contains no yeast).

Beyond that, I usually shop twice a week for fresh produce, and have a selection of fresh veggies and low-sugar fruits in the house at all times. You won't find them all at once, of course, but I often have baby field greens, dark lettuces, broccoli, spinach, kale, red and green cabbage, celery, carrots, cucumber, cauliflower, squash, courgettes, beetroots, radishes, avocado, tomato, bell peppers, chard, asparagus, green beans, leeks, and aubergine—basically, a variety of the foods eaten freely on this program. I always have some fresh herbs, especially parsley, basil, and coriander, in my kitchen, and usually some fresh salsa (made with lemon or lime juice, never vinegar).

There are many, many more kinds of produce, of course, not mentioned here. Your selection should reflect your tastes, and your family's, as well as the recipes you use most often. Your kitchen won't look exactly like mine, but once it is well stocked, you'll find it is easy and natural to follow this way of eating. (Remember to read labels carefully to avoid hidden harmful foods, especially citric acid, sweeteners, artificial sweeteners, yeast, vinegar, peanuts, and corn syrup.) There's no need to go out and buy every single thing on this list at once. Start with a few items from each category and build up as you go along. Let it be an adventure to create a selection of alkalizing food choices for you and your loved ones!

THE RIGHT TOOLS

You can make a great alkalizing meal with just one good knife—and plenty of time. However, as any good carpenter knows, the proper tools do the job faster, easier, and with optimum results. Here is a quick look at the things I (Shelley) find are indispensable in my kitchen.

Good knives are a must for cleaning, trimming, cutting, and chopping your veggies. I got along quite well with a three-piece starter set for years before investing in a larger set. I use them many times every day, and simply could not get by without them. If you've had poor quality or dull knives, you'll be amazed by the difference a high-quality, properly sharpened knife can make. I'm a fan of Cutco knives, which come with a lifetime guarantee and hardly ever need sharpening.

A **food processor** will cut your chopping, blending, and mixing time by as much as 90 percent, especially when you are preparing food for a crowd. I use a Cuisinart. I started with standard seven-cup (five pints/two litres) size, which is fine for everyday needs. Look for one that comes with both sharp and soft-edged S-blades, as well as shredding and slicing wheels. The sharp S-blade is useful for mixing, mincing, fine and coarse chopping, blending foods like hummus, emulsifying things like salad dressing, and grinding dry ingredients like nuts, seeds, grains, and dried tortillas into powders. I use the soft-edged S-blade to mix the dough for tortillas. And I slice and shred all kinds of things—which I especially love because it makes it easy to make beautiful salads.

A **blender** is useful, too, for mixing, blending, and grinding. I use the Vita-Mix®. It has a strong motor and a good variety of options for speed control (as well as "reverse"). The unique feature, and one I use a lot, is that if you leave it running longer, the friction it creates warms the food—so recipes like raw soups can be worked up very fast and then served immediately straight from the blender. (Of course, you can also serve them cold—just blend for a shorter period of time.)

My **rice cooker** is almost always on the counter, full of freshly steamed brown rice, buckwheat, or other grain for my family to help themselves to any time during the day. I use the Zojirushi® brand, which will also cook legumes.

Sometimes I start it up before I go to bed at night so we'll have warm rice for breakfast.

I also could not do without (well, would never *want* to do without) my **salad spinner.** I use it all the time to wash and dry greens in a jif.

I also like my **mandolin** for doing extra-fancy cuts on veggies. Presentation is a key part of how you experience a meal, and this is a simple way to make dishes more elegant and visually appealing.

I'm a fan, too, of a small hand machine called a **Saladacco** (see Resources) that can make angel-hair "pasta" out of vegetables such as squash. (I use yellow summer squash and zucchini, but any squash would work.) It also makes beautiful ribbon cuts that look like Hawaiian leis on such vegetables as beetroots and carrots.

A set of **cups** is useful in every kitchen. In the USA, cups are used to measure the volume of ingredients—liquid or otherwise. Equivalent fluid measurements are: 1 cup = 250 ml/9 fluid oz.; $\frac{1}{2}$ cup = 125 ml. 5 fluid oz.; $\frac{1}{3}$ cup = 80 ml./3 fluid oz.; $\frac{1}{4}$ cup = 60 ml./2 fluid oz. Cups are available in the UK from good cookware shops including Lakeland Limited (see Resources).

Once you are committed to preparing mostly raw foods for your meals, you can really get into gadgets like these. Invest in them as you can, to make preparing your food simpler—and more fun (and sometimes prettier). Experiment, and enjoy.

SPROUTING

Sprouts, with their tremendous regenerating properties, are incredibly nutritious and alkalizing. All the wide variety of vitamins and minerals in a seed or grain explode when sprouted. Sprouting also takes starches, proteins, and hormonal agents in the seed and turns them into very alkalizing, easily assimilated, predigested proteins and subtle vegetable sugars. Finally, the phytochemicals that fight cancer appear in plants just as soon as they sprout. I think a person could live on sprouts alone with no (physical) problem at all.

So, I keep plenty of fresh sprouts around all the time—I grow them. No green fingers necessary. Really. The process is simple. In no time at all, a pinch of seeds almost magically transforms into many times their original weight in fresh produce, some in as little as two days.

Start with organically produced seeds. You can store seeds for sprouting for long periods of time (up to ten years if they are unopened, and one to two years if opened but kept dry and cool), so stock up and keep a variety on hand. Some of the easiest sprouts to grow are alfalfa, mung bean, chickpea, green lentil, sesame, sunflower, buckwheat, and wheat. The Sprouting Guide table provides basic directions for many common sprouts.

I soak the seeds overnight in purified water, until they're plump, then pour them into sprouting trays, let them drain, and place the trays in a dark, warm cupboard, then rinse them twice a day. Actually, I keep the trays where we keep our water, so that when I get my first drink of the morning and my last one at night, I give the sprouts a drink, too, rinsing and draining them.

In two or three days, most sprouts will be ready to eat (check Sprouting Guide). Sprouts should be crisp and slightly sweet—never sour. If they are souring, as the shop-bought ones often are, they have gone too long, and you should start again. Sprouts should never be browning or at all

SPROUTING GUIDE

Seed	Quantity	Soak time (hours)	Rinse/ Drain (times daily)	Time to harvest (days)	Height to harvest (inches)
Alfalfa	2 tbs.	6 to 8	2	3 to 6	1 to 2
Chickpea	1 cup	16	2 to 3	3 to 6	$1/8$ to 1
Chinese cabbage	1 cup	6 to 8	2	3 to 4	$1/2$ to 1
Fenugreek	1 cup	6 to 8	2 to 3	3 to 4	$1/2$ to 1
Lentil	1 cup	8 to 12	2 to 3	2 to 4	$1/2$ to 1
Mung bean	$1/2$ cup	8 to 12	2 to 3	2 to 4	$1/2$ to 1
Peas	$1/2$ or 1 cup	8 to 12	2 to 3	2 to 3	$1/2$ to 1
Radish	2 tbs. or 1 cup	6 to 8	2	3 to 4	$1/2$ to 1
Red clover	2 tbs.	8	2	3 to 6	$1/2$ to 2
Sesame	$1/4$ cup	8	2	1 to 3	0 to 1
Soybean	$1/2$ or 1 cup	16	3	3 to 5	$1/2$ to 1
Sunflower, hulled	$1/2$ or 1 cup	6 to 8	2	1 to 2	0 to $1/2$

slimy. Store sprouts in the refrigerator in a glass jar or sealed plastic bag. They will keep for about a week.

You can also grow sprouts using two-pint jars with mesh rubberbanded over the top, or special sprouting jars with drainage lids, instead of the trays. Go with whichever method you find simplest!

I recommend kits to help get you started with sprouting (see Resources). Many health food shops and suppliers found on the Internet carry sprouting supplies, including seeds ready for sprouting.

WAYS TO USE SPROUTS

Sprouts are terrific raw, stir fried, or steam fried, on their own, or in sandwiches or salads, or sprinkled over soups. Try them all different ways—the more sprouts you get, the better. Everyone should make sprouts a part of their daily diet.

Almost any sprouts are good in salads, and I (Shelley) use them all the time in sandwiches and wraps, soups and juices, steamed dishes, and casseroles. They also make a great snack. Here are my suggestions:

Salad: alfalfa, chickpea, Chinese cabbage, fenugreek, lentil, mung bean, peas, radish, red clover, sunflower
Sandwich: alfalfa, radish, red clover
Juice: alfalfa, Chinese cabbage, radish, red clover
Soup: chickpea, lentil, mung bean, peas, soybean
Casseroles: chickpea, sesame, soybean
Snacks: fenugreek, sesame, sunflower
Steamed: lentil, mung bean, peas, soybean

This is just to get you started! Use your imagination, experiment, follow your tastebuds. Just keep eating sprouts!

DEHYDRATING

Using a dehydrator is a wonderful way to serve your food warm, but not cooked. It also makes it easy to keep fresh veggies in your pantry (once they are dried, fresh vegetables stored in an airtight bag or container will keep for at least a year in a dry, cool place). Food dehydrators are also useful for warming pâtés and loaf-type recipes before serving.

Dehydrated vegetables and nuts make great snacks and terrific garnishes. Enjoy them on their own, or with a favorite dip or pâté. They are very pretty sprinkled over soups, add texture to salads, and can accent any plate nicely.

Dehydrating most vegetables couldn't be simpler. You just clean and slice them (about a quarter of an inch thick), marinate if desired, drain, and place on clean drying racks in your dehydrator. Dehydrate until all water is out of the vegetables and they are crisp.

You can do this with just about any vegetable. I especially like carrots, tomatoes, onions, celery, and bell peppers. For root vegetables such as winter squash, carrots, and yams, I like to marinate them for up to an hour in Bragg™ Liquid Aminos, garlic, ginger, and spices.

Try dehydrating nuts, too. Start with soaked nuts. Marinate, as above, in a shallow bowl for one to 12 hours. Drain well, place in dehydrator, and dry until crunchy. Store in an airtight container in the fridge.

I use an Excalibur® brand dehydrator (see Resources) because I like the way the air circulates throughout each tray, rather than just coming from the bottom up. The flexible, teflex liners also make it easy to make batter-type recipes, such as Dehydrated Linseed Chips (see page 378), and to lift off foods after they have dried.

TIPS AND TRICKS

Here are some of my favorite shortcuts for making and arranging alkaline meals quickly and easily:

- Keep a huge salad in the fridge at all times. I make one that will last about three days and fill it with such goodies as spinach, red onions, pine nuts, tofu cubes, shredded carrot and beet, radishes, and sunflower seed sprouts. Then I can grab a quick salad or fill a wrap quickly. It is also good to have on hand when kids come home from school with that ravenous appetite.
- Make enough of your favorite salad dressings to last all week.
- Use prepared spice combinations. I keep a selection in the pantry to add interest and variety to whatever dish I'm making.
- Keep a bowl of soaked almonds in the fridge. They are great for a sweet, crunchy snack, and on salads instead of croutons. They are also good for whipping up some nut "milk" in a hurry. Just cover raw nuts with plenty of water, soak overnight, and change the water daily. They'll keep for about three days.
- Mix up batches of your favorite spread, such as pesto or hummus, for use as a dip for raw veggies, a topping for steamed veggies, to spread on crackers, or to tuck into

wraps. You can find some good ones at your health food shop or supplier, including dairyless pesto, although you must always read the labels carefully so you know exactly what you're buying.

- Double or triple a cooked recipe and freeze for future meals or quick snacks.
- Keep lemons and limes on hand to use as a vinegar substitute and to squeeze into your drinking water all day long. I use lemons somewhere in almost every meal I serve.
- Take a few packages of sprouted wheat tortillas and set them out to dry on your kitchen counter overnight. Or bake them in a low-heat oven, about 300°F/150°C/gas 2, for fifteen to twenty minutes until crisp but not browned. They should break easily, like a cracker. Grind them in your food processor or blender until they are like flour, to use when a recipe calls for bread crumbs or white flour. Store in an airtight container. They will keep for a few weeks in a cool, dry place (less in humid weather).
- Use your freezer to advantage. Store serving-sized packages of nuts, herbs, and even good oils such as linseed, so you'll have them when you need them (and they'll stay fresher, and defrost quickly). I keep a fresh ginger root in the freezer, and it is then easy to grate when a recipe calls for it. My favorite use for it, though, is to make fresh lemon ginger "tea" after a nice meal.
- Keep a rice cooker on the counter with a fresh batch of steamed legumes or grains for that 20 to 30 percent of your diet.
- Learn to do your own sprouting and keep fresh sprouts on hand for a great snack—or a nutritional booster to any meal.

- Be prepared for the desire for crunchy snacks by keeping healthy options easily available, such as baked sprouted tortillas, raw almonds (best soaked), raw veggies (one of our favorites is slightly sweet arrowroot sticks, which are good anyway, but can also help during sugar cravings), and baked tofu.
- Stock up on good water—and drink up! When your meals are 80 percent alkaline, you may not feel a need to drink with meals, since most vegetables are already 70 to 90 percent water. But in between meals, drinking a lot of pure water is one of the very best things you can do for yourself.

Part III

THE
RECIPES

CHOOSING THE RIGHT RECIPE

I'll let you in on a secret: Before I understood about how foods heal, I never really enjoyed cooking. It was just another time-consuming chore that made a mess—and had to be done all over again just about as soon as you got it all cleaned up. I felt as if I was stuck in the kitchen when I would much rather have been out there in the rest of the world. But now that I realize how great an impact what we eat has on our health and well-being, I love to create healing meals. And I treasure the tangible nurturing that feeding my family represents. I still want the process to be efficient, though! I like to spend a little time preparing a meal—and a lot of time relaxing and enjoying every mouthful (and enjoying the company of my family, which is also part of the healing power, if you ask me!).

I also want to create something I'll want to eat—and that my family will enjoy, too. So this section is composed of the recipes I use that keep us all happy. I'm not too long in the kitchen, we all eat healthfully—and we love what we eat. We eat heartily, too, I might add, since some people seem to think at first that this way of eating is "for the birds," or "rabbit food." All of these dishes celebrate variety, texture, and out-of-this-world flavors at every meal. It is never boring. (All this and they are easy on the wallet, too. It doesn't cost any more to buy organic ingredients than it did to pay for meat, cheese, sweets, beer, and so on.) This program, far from being any kind of deprivation, is a gift we give ourselves.

A lot of people find that their palates are jaded after so much artificial nonfood, and so much artificially stimulating hard sugar and salty flavor, and they are insensitive, at first, to the subtleties of whole, natural foods. If that's you, never fear: As you adjust to eating in this 80/20 ratio, you'll find your tastebuds will awaken to the glorious sensations you are treating them to. You'll also leave behind that strange willingness to self-destruct for pleasure I've seen in so many people for whom food is one of life's great entertainments. Some folks live to eat, though it should rightly be the other way around. On this program, food regains its proper perspective—we eat to live—without losing its tremendous sensual appeal. A good meal can still be tremendously enjoyable and provide deep satisfaction. It doesn't have to be killing you to do so.

Many of these recipes are not for use during a Cleanse (though many of the juices and soups *are*). And many are not for the weeks immediately following a Cleanse. And there are many included that are meant for use once you've completely rebalanced your system. For as long as you are facing the challenge of symptoms, the simpler, "rawer" choices will be best for you. The majority of them are good just about any time (though perhaps not on the Cleanse), and should always be the bulk of your meals, no matter what stage you're in.

So keep the guidelines you've learned in this book in mind as you select the recipes that are right for you, wherever you are in the program. You can also look for the lists I've included of suggested dishes for different phases to guide you. Take a look, too, at the ideas for breakfasts, healthy snacks, and really quick meals. Once you are familiar with the basics, this program should be not just healthy, but simple and easy—as well as enjoyable and delicious!

RECIPES FOR THE CLEANSE

Basic Green Vegetable Juice
Garden Green Drink
All-Vegetable Cocktail
Blood Builder (juice)
Vegetable/Grass Drink*
Basic Green Drink*
Green Power Cocktail*
Spring Green Drink*
Potassium Special (juice)*
Insulin Generator (juice)*
High Vitamin C & E Drink*
Skin Cleanse (juice)*
Anticancer Soup
AsparaZincado Soup
Broccoli/Cauliflower Soup
Celery Soup
Celery/Cauliflower Soup
Creamy or Crunchy Broccoli Soup
Creamy Vegetable Soup
Gazpacho
Madrid Gazpacho
Green Raw Soup
Healing Soup
Mock Split Pea Soup
Popeye Soup
Sweet Pepper Consommé

Vegetable Minestrone
Almond Milk (fresh raw, not processed)
Alfalfa Sprout Salad**
Alkalizing/Energizing Cucumber Salad**
Bean Sprout Salad**
Broccoli Salad**
Cauliflower Toss**
Colorful Cabbage (take carrot out if juicing)**
Courgette Toss**
Potassium Salad (take carrot out if juicing)**
Rainbow Salad (take out beetroot, carrot, arrowroot if juicing)**
Spinach Salad**
Sprouted Lentil Salad**
Wheat Sprout Salad (take carrot out if juicing)**

*Omit carrot and beetroot during the Cleanse. See Chapter 8 for more information on using juices.

**Put in a blender and eat as soup during a Cleanse. Or put ingredients through a juicer and make into a fresh juice—leaving out any carrots or beetroots called for when you are on a Cleanse.

Breakfast Ideas

Salad. Especially when the weather's hot, a nice salad with lots of sprouts, garnished with linseed oil, lemon, and Bragg's (or your favorite dressing), makes a refreshing breakfast.

Buckwheat Cereal with almond milk is quick and filling on a cold morning. Buckwheat groats, cracked or cream of buckwheat are available in most health food shops.

Sprouted Cereal (see page 382).

Soup. I love to warm up with soup for breakfast on a chilly morning. On warm days, I choose cool soups! (AsparaZincado Soup [see page 272], Popeye Soup [see page 274], and Creamy or Crunchy Broccoli Soup [see page 270] are particularly satisfying starts to the day.) Hearty soups and stews course through the body creating a good glow that lasts for hours.

Vegetable Juice. A glass of fresh juice is a perfect light—but incredibly nutritious—eye opener.

Zippy Breakfast. This is, hands down, the favorite around our house. It is a hearty choice. I usually don't feel hungry again until mid-afternoon after a Zippy Breakfast! It is basically any healthy warm whole grain, topped with avocado and tomato plus other veggie garnishes you like, drizzled with good oil, lemon or lime juice, and Bragg's, and sprinkled with The Zip (Spice Hunter brand) or other zesty seasoning (see Zippy Breakfast on page 328).

Wraps. A bunch of fresh or steamed vegetables, soaked seeds or nuts, and a few sun-dried tomatoes in a sprouted wheat tortilla (or just about any other combination you dream up) is as portable as it is quick and delicious.

Casserole de Cauliflower. This dish is good any time (see page 341), but particularly satisfying at breakfast.

Steamed Broccoli. Steam florets lightly for five minutes. Add chopped onion and/or another green vegetable, stir in some Basic Salad Dressing (see page 268), and top with soaked almonds or hazelnuts.

Quick Meal Ideas

Salad. Keep clean, dried greens, and a few salad dressings, in your refrigerator, so all you have to do is throw on a couple of handfuls of whatever chopped-up goodies catch your fancy that day, and you're ready to eat.

Wraps. Start with a burrito-size sprouted wheat tortilla, spread with hummus or another of your favorite spreads. Add some sprouts, beans, avocado, tomato, and/or any other veggies you have on hand or are in the mood for. Spray with Bragg Liquid Aminos and/or top with fresh salsa or pesto. That's a wrap!

Soup. Pick a quick recipe, such as Madrid Gazpacho (see page 284) to whip up when you are ready. Or make a big batch over the weekend to see you through the week. With raw soups served chilled you don't even have to take the time to warm them!

Any Time Recipes

These recipes require no cooking (they are raw) and are 90 to 100 percent alkalizing. They can be used any time, including the early weeks of the program, immediately following a Cleanse, or when you are dealing with any specific imbalances or serious symptoms. Once you are in balance, these are all perfect for that 70 to 80 percent of your meal that is alkaline.

Alfalfa Sprout Salad
Alkalizing/Energizing Cucumber Salad
AvoRado AvoCado Topping
Bean Sprout Salad
Broccoli Salad
Cauliflower Toss
Colorful Cabbage
Courgette Toss

Pretty Ribbon Quiche
Potassium Salad
Rainbow Salad
Spinach Salad
Sprouted Lentil Salad
Wheat Sprout Salad

Maintenance Recipes

Once you feel balanced and are no longer dealing with symptoms, you'll have a larger group of recipes to explore while maintaining that 70/30 or 80/20 split to your meals—with these being in that 20 to 30 percent. For example, you might have a big salad with a side of baked Tofu Italian Mock Meatballs.

These choices add more texture and all-important variety to your repertoire. Some of these recipes are warmed, processed, or cooked, and so require more time to digest. These are not good choices in the early weeks after a Cleanse, or when experiencing an acute illness or imbalance in the body.

These meals are also good choices for someone who is transitioning slowly into an alkaline way of eating, eliminating acid foods such as meat, dairy, sugar, and fruit gradually.

Alexandra's Favorite Pasta
Almond/Carrot/Ginger-Stuffed Courgette
Autumn Curry Crêpes with Curried Veggie Filling
Baked Falafel Fritters
Blackened Herbed Fillets
Cabbage Stuffed Vegetables

Cajun-Style Red Beans and Brown Rice
Casserole de Cauliflower
Cold Tofu Pockets
Doc Broc Casserole
Edamame (Soybean) Patties
Ginger-Almond Paste Topping
Green Chili Tofu Pita
Hearty Harvest Casserole
Maren's Tortilla Pockets
Millet/Buckwheat Oven Cakes
Millet Yam Hash Browns
Nepal Vegetable Curry
Nutty Mock Meat Loaf
Popeye Mousse Pie
Sautéd Edamame (Soybean) Vegetable Soup
Shelley Beans
Shelley's Super Tortillas
Shelley's Super Wraps
Sprouted Bean Casserole
Stuffed Acorn Squash
Stuffed Cabbage Rolls
Sunrise Asian Salad
Three-Bean Salad
Tofu Italian Mock Meatballs
Tofu Patties
Tofu Spinach Quiche
Tofu Stew
Wild Yam Soba Noodles with Kale and Spicy Pine Nuts
Vegetable Steam Fry
Zippy Breakfast

80 or 20?

These recipes fit into either side of that 80/20 or 70/30 split, depending on whether they are raw or cooked. When they are cooked, they move from the 80 to the 20 (or the 70 to the 30).

Anticancer Soup
AsparaZincado Soup
Broccoli/Cauliflower Soup
Celery Soup
Celery/Cauliflower Soup
Chunky Veggie Soup
Creamy or Crunchy Broccoli Soup
Creamy Vegetable Soup
Curried Squash Dhal
Gazpacho
Green Raw Soup
Healing Soup
Mock Split Pea Soup
Popeye Soup
Roasted Butternut/Celery Soup with Caramelized Onions
Special Carrot Soup
Sweet Pepper Consommé
Thick Purée of White Bean Soup
Vegetable Minestrone
Veggie Borscht

Healthy Snacks

These healthy snacks satisfy the need for crunch and munch that kids of all ages have, and they are perfect when you just need a quick bite or are on the go. Or maybe you have one of those kids in your house who seems to live on snacks? These are also useful in transitioning, and some help during sugar cravings. Some of these travel really well, so you can make sure to continually nourish yourself no matter where you are or what you are doing.

Of course, you can always grab a simple snack like raw almond butter on a brown rice cake or a sprouted wheat tortilla. But if you are in the mood to create, or just want something different, or are in the mood for a treat, try:

All-Vegetable Cocktail
Almond Pâté
Ashley's Vegetable Nori Roll-Ups
Avocado/Tomato Snack
Camper's Bread
Chickpea Spread
Chilled Cucumber Refresher
Courgette Italian Style
Crispy Buckwheat Groats
Crispy Radish Filling
Curried Veggie Filling
Dehydrated Linseed Chips
Dried veggies and nuts for crunch and munch
Essene Bread
Fresh Cucumber Dills
Fresh Spinach Filling
Garden Variety Filling
Great Olé Guacamole
Hearty Nut Filling
Kale with Egyptian Garlic Sauce
Leprechaun Surprise Dip
Mexicali Rice

Mock Pumpkin Pie
Okra and Tomatoes Creole
Raw Pecan Pâté
Refried Beans
Spiced Green Beans
Spiced Winter Squash
Spicy Pecan Croutons
Sprouted Cereal
Sprouted Wheat Bread
Steam-Fried Sprouts
Tofu Salad Spread
Veggie Crunch Stix and Crackers
Yummus Hummus
Zippy Garbanzo Spread

SALADS

Salads are my favorite meal. I'll often have one for break-fast—actually, almost every time I sit down to eat. Traditionally, a salad has been a token gesture compared to the "real" meal, but we need to rethink this. Salads *are* a main course. Other vegan dishes, including grains, soups, tortillas, pâtés, and other warmed and cooked foods, should be the complement to the salad, rather than the other way around.

Fortunately, making a great salad couldn't be easier. Even complete kitchen novices can shine here, from day one. And the variety is limited only by availability of ingredients and your imagination. True, it may take a bit longer than push-ing a few buttons on the microwave to zap a meal. But then a big salad, stored airtight in the fridge, will stay fresh for about three days. (Use a salad spinner to remove excess water and thoroughly dry your greens so they will stay crisp.) Keep clean, dry, dark, leafy lettuces and spinach in your refrigerator in a covered container with a paper towel in the bottom.

Get creative with salads. Showcase the simplicity of vegeta-bles with just a few ingredients, or show off the complex interplay of a whole bunch of additions. Serve them with small piles of each ingredient on its own turf, or tossed all together hodge-podge. Mix up how you prepare the veggies—minced, diced, sliced, shredded, chopped, and so on—or keep all the textures similar. Make them monochromatic (green being a popular color scheme) or brightly jeweled rainbows. Keep them light, or make them extra hearty by adding tofu cubes (or baked or marinated tofu cubes), pine nuts, soaked almonds, sprouts, dehydrated grains (such as Crispy Buckwheat Groats, page 378), or pieces of baked warm falafel. Have fun, explore the salad horizons—and eat well.

Alkalizing/Energizing Cucumber Salad

Serves 3

Cucumber is one of the most alkalizing and energizing foods that you can eat. It is considered to have a purifying effect on the digestive system and is very beneficial to the hair and skin. (For a refreshing lift, lie down with a cucumber slice over each eye for a few minutes or rub a slice over your face after cleansing to tone and purify your skin.)

2 cups cucumbers, chopped
2 tbs. parsley, chopped
1 tbs. lemon juice
1 tbs. linseed oil or olive oil
⅓ cup finely chopped peppermint

In a small serving bowl, combine the cucumbers, parsley, lemon juice, oil, and mint. Toss together. Chill for several hours or overnight. Toss again before serving.

Rainbow Salad

SERVES 8–12

I love the colors of vegetables! Presentation of a beautifully arranged alkalizing meal can be an art. Besides, eating a rainbow of colored foods supports the balance of the energy of the body. This salad is my basic salad recipe that I make every week. The grating of the vegetables exposes their natural sweetness.

1–2 heads of green-leaf lettuce, washed, dried, and ripped into bite-size pieces
1–2 heads of red-leaf lettuce, washed, dried, and ripped into bite-size pieces
1 pack of prewashed baby leaf organic spinach
1 head green cabbage, shredded
$\frac{1}{2}$–1 head red cabbage, grated
3–4 beets, grated
4–5 carrots, grated
2–3 summer (yellow) squash or courgettes, grated, *or* $\frac{1}{4}$ butternut squash, grated
$\frac{1}{3}$–$\frac{1}{2}$ large arrowroot, grated
1 each red, yellow, and orange bell peppers, sliced
1–2 cucumbers, sliced
1–2 packs (approximately 8 oz, 200g.) sunflower seed sprouts, *or* a mix of sprouts, or the sprouts of your choice
1 lb. fresh green peas from the pod
1–2 tbs. salad dressing of your choice *per serving*

Fill a large salad bowl with lettuces. (You can substitute a packaged salad mix: Get it as fresh as possible, and choose an organic one.) Arrange grated vegetables on top of the greens, starting with the deeper, darker colors on the outside and working into the lighter colors on the inside to create a rainbow effect. Place sliced peppers and cucumbers on top. Sprinkle with sprouts and peas. Top with dressing, or pass it at the table. Choose a dressing from any of the recipes in this book, or any you buy that fits on this program. Or just sprinkle with a healthy oil, Bragg™ Liquid Aminos, fresh lemon juice, and spices to taste. I hope once you've tried this, you'll feel free to improvise your own ingredients, proportions, and arrangement.

Sprouted Lentil Salad

SERVES 4

This salad is hearty and also works well as a filling for halved bell peppers.

1 tsp. linseed oil, Udo's Choice® Oil, or Omega
   Nutrition/Essential Balance Oil
1 tbs. lemon juice
1 tsp. Bragg™ Liquid Aminos
1 clove garlic, minced
Pinch of Zip (Spice Hunter) or alternative
1 tsp. Curry Seasoning (Spice Hunter) or curry powder
2 cups sprouted lentils
1/2 cup chopped onions

In a small bowl, mix the oil, lemon juice, Liquid Aminos, garlic, Zip, and curry powder. In a separate mixing bowl, combine the lentils and onions and toss.

   Optional: Add some sprouted chickpeas, too!

Spinach Salad I

SERVES 2–3

1 head spinach
$^1/_2$ cup cauliflower, cut in small pieces
2 stalks celery, chopped
6 radishes, chopped

In a large bowl, combine ingredients and toss well. Top with Essential Dressing (see page 268).

Spinach Salad II

SERVES 2–3

1 head spinach
$^1/_2$ cup cauliflower, cut in small pieces
2 stalks celery, chopped
6 radishes, chopped
2 shallots, chopped (or 1 small red onion)
$^1/_2$ cup chopped basil
2 red peppers, chopped
4 tbs. pine nuts

In a large bowl, combine all ingredients and toss well. Top with Essential Dressing (see page 268).

Bean Sprout Salad

SERVES 4

$^1/_4$ cup linseed oil
2 tsp. fresh lemon juice
2 tbs. Bragg™ Liquid Aminos
$^1/_2$ tsp. freshly ground pepper
1 crushed clove garlic
2 tbs. sesame seeds (soaked overnight)
2 cups fresh bean sprouts
$^1/_4$ cup finely chopped pimento
$^1/_4$ cup finely chopped green onion

For dressing, combine oil, lemon juice, Liquid Aminos, pepper, garlic, and sesame seeds in blender, and purée.

Rinse bean sprouts in cold water and drain. In a bowl combine bean sprouts, pimento, and onion. Toss lightly in puréed dressing.

Three-Bean Salad

SERVES 2

Mix in large bowl:

6 oz./175 g. steamed fresh green beans
6 oz./175 g. steamed fresh white (haricot) beans
6 oz./175 g. cooked red kidney beans, drained
$^1/_2$ cup green onion, chopped
$^1/_4$ cup snipped fresh parsley

Dressing:
¼ **cup linseed oil, Udo's Oil, or Omega**
   Nutrition/Essential Balance Oil
⅛ **cup Bragg™ Liquid Aminos**
2 **cloves garlic, crushed**
½ **tsp. Italian seasoning (Spice Hunter or other)**

Pour dressing over bean mixture, refrigerate for two hours. Just before serving, remove bean mixture with slotted spoon onto lettuce bed.

Potassium Salad

Serves 4

1 **head cabbage (green or white), shredded or finely**
   chopped
¾ **cup chopped parsley**
3 **grated carrots**
1 **avocado, sliced into bite-size pieces (optional—it adds**
   even more potassium)

Dressing:
⅔ **cup linseed oil**
¼ **cup Bragg™ Liquid Aminos**
Dulse, garlic powder, onion powder

Combine vegetables in a large bowl. Mix well. Toss with dressing, to taste.

Alfalfa Sprout Salad

Serves 6

3 cups alfalfa sprouts
3 cups chopped summer squash
2 red peppers, diced
2 chopped green onions
1/4 cup chopped red onion

Dressing:
1 cup linseed oil
Juice of 1 fresh lemon or lime
1 tsp. Real Salt™ or alternative
1–2 tsp. seasoning blends (optional), such as Italian or
    Mexican (I use Spice Hunter brand)

Combine vegetables in a large bowl. Toss with dressing to taste.

Wheat Sprout Salad

Serves 6

3 cups fresh sprouts
1 cup grated carrots
3/4 cup minced onion
3 tbs. linseed oil, Udo's Oil, or Essential Balance/Omega
    Nutrition Oil
1 1/2 tbs. fresh lemon juice
Paprika

Mix all ingredients except paprika. Sprinkle with paprika.
Serve on a bed of lettuce.

Broccoli Salad

SERVES 2–4

1 head broccoli
1 cup diced celery
4 chopped spring onions
1 large chopped red onion
$\frac{1}{3}$ cup Parsley Dressing or Herbed Salad Dressing (see pages 269 and 264)

Cut raw broccoli into bite-size pieces. Mix all ingredients, chill for one hour.

Colorful Cabbage

SERVES 4

Cabbage is considered one of the most powerful therapeutic foods in the world. Many studies have linked eating cabbage with a reduction of cancer, especially colon cancer. Also, cabbage juice has been proven to help heal stomach ulcers and prevent stomach cancer.

2 cups red cabbage, thinly sliced
2 cups green cabbage, thinly sliced
1 carrot, grated
1 red pepper, slivered
1 yellow pepper, slivered
1 green pepper, slivered
1 orange pepper, slivered
4 tbs. spring onions, chopped
4 tbs. parsley, minced
$\frac{1}{4}$ cup lemon juice

3 tbs. water
1 tbs. oil (extra virgin olive, linseed, or Udo's Choice®)
1–2 tsp. dried red chili pepper
Dash of Bragg™ Liquid Aminos

In a bowl, combine all ingredients. Toss thoroughly and let the flavors mix for at least half an hour before serving.

Zucchini Toss

SERVES 4

1 medium head red-leaf lettuce
1 small romaine lettuce
2 medium courgettes, thinly sliced
1 cup sliced radishes
3 green onions, sliced

Dressing:
¼ cup linseed oil
2 tbs. Real Salt™ or alternative
Garlic clove(s) to taste, crushed
¼ tsp. dried tarragon leaves

Combine all vegetables in a large bowl. Toss with dressing.

Cauliflower Toss

SERVES 4

½ medium bunch romaine lettuce, torn
½ small head cauliflower (broken into florets—2 cups)
¼ cup radishes, sliced

Dressing:
¼ cup linseed oil
1 green onion, sliced
¼ tbs. dried dill weed
Vegetized or sea salt
Freshly ground pepper

Layer half each of lettuce and cauliflower in salad bowl. Top with radishes and remaining lettuce and cauliflower. Combine dressing ingredients, mix, and pour over salad.

Fresh Cucumber Dills

SERVES 6

2 large cucumbers, peeled and thinly sliced

Stir Together:
2 tbs. fresh dill weed
1 tbs. fresh lemon juice
3 tbs. distilled water
½ tsp. Real Salt™ or alternative
Dash of cayenne pepper

Drain cucumbers well, add dressing. Stir in well. Cover and chill overnight.

SALAD DRESSINGS

You can thicken dressings to your liking with ground linseeds, ground psyllium seed powder, agar agar, or kuzu root, all of which you can find from your local health food supplier.

Cinnamint Dressing

SERVES 4

This is a fresh-tasting dressing that is not salty.

5 tbs. carrot juice
½ cup olive oil
⅓ cup lemon juice
1 tsp. Orange Ginger Pepper blend (Spice Hunter) or alternative
½ tsp. lemon pepper
½ tsp. cinnamon
⅛ tsp. paprika
1 tbs. fresh mint, finely chopped

Put all ingredients except mint in a food processor or blender. Blend until smooth. Stir in mint.

Soy Cucumber Dressing

SERVES 4

A subtle refreshing dressing.

2–3 tsp. carrot juice
1 large cucumber (I prefer peeled and seeded)
$^1/_2$ red bell pepper
$^1/_2$ small onion
1 cup soy milk
1 tsp. dried basil (or 2 tsp. fresh)
1 tbs. Bragg™ Liquid Aminos, Real Salt™ or alternative, to
   taste

Blend all ingredients in food processor or blender until smooth.

Herbed Salad Dressing

SERVES 2

1 tsp. dry mustard
1 tsp. fresh parsley
1 tsp. dill weed
$^1/_2$ tsp. Real Salt™ or alternative
$^1/_4$ tsp. tarragon
$^1/_4$ tsp. ground black pepper
$^1/_8$ tsp. thyme
$^1/_3$ cup preferred oil (virgin olive, grape seed, Udo's,
   linseed, or Essential Balance)
Pinch of oregano

Combine all ingredients.

Spicy Asian Dressing

Serves 4–6

This dressing is easily whipped up in the food processor and gives your salads a wonderful Asian zing. If you tend not to like too much spice, then use half the amounts of spices listed.

$^1\!/_3$ **cup plus 1 tbs. sesame tahini**
$^1\!/_2$ **cup water**
$^1\!/_2$ **cup Bragg™ Liquid Aminos**
2 tsp. dried onions
6 tbs. linseed oil
$^1\!/_2$**–1 tsp. chicory root powder (sweetener)**
3 tbs. grated ginger root
$^1\!/_2$ **tsp. Chinese 5 spice seasoning**
$^1\!/_8$**–**$^1\!/_4$ **tsp. cayenne pepper or Hot Zip (Spice Hunter)**
$^1\!/_2$ **tsp. cumin**
Optional: Add water to thin if needed

Place all ingredients in food processor and process until smooth and well mixed. This dressing will last several days in the refrigerator.

Curried Carrot Almond Dressing

Serves 4

This dressing is simple, fast, and full of great flavor. I have found that it is best if you first blanch the soaked almonds to remove their skins. Since this dressing has fresh carrot juice in it, it is highly perishable, so make only enough to use up in one or two days.

$^{1}/_{2}$ cup almonds, soaked and blanched with skins removed
    (a rubber garlic cylinder roller takes the skins off fast)
1 cup fresh carrot juice
$^{1}/_{3}$ to $^{1}/_{2}$ tsp. curry powder
$^{1}/_{2}$ tsp. dried onion
Squirt of Bragg™ Liquid Aminos, to taste
Dash of lemon or lime juice to taste
Optional: 1 clove fresh or 2 cloves roasted garlic

Put all ingredients in a blender and blend at high speed
until smooth. If you want this for more of a dip, then use
more almonds and less carrot juice and process to desired
thickness.

Tahini Dressing/Dip

SERVES 4

1 cup raw tahini
1 tsp. dried parsley or $^{1}/_{2}$ cup fresh parsley
1 tbs. dried onion or 1 small onion, chopped
2–3 sun-dried tomatoes
$^{1}/_{2}$ cucumber, peeled and chopped
1 fresh tomato, chopped
2 tbs. lemon or lime juice (I use more)
1 tsp. fresh coriander, finely chopped
$^{1}/_{2}$ tsp. cumin
1–3 cloves garlic or 4–6 cloves roasted garlic
Shot of Bragg™ Liquid Aminos, to taste
1 tsp. Real Salt™ or alternative
Pinch of cayenne or Zip (Spice Hunter)

Put all ingredients in a food processor and blend until smooth. Thin with water/oil for dressing, or let it set in the fridge for a dip.

Wowie Zowie Almond Butter Dressing

SERVES 6

This is a rich, sweet dressing that tastes great on salads, over rice, or on top of any steamed veggies. Children will especially like this dressing because of its creamy nut butter taste. Add additional water to thin if desired.

1 cup raw almond butter
$^1/_2$–1 cup water
Juice of 1 lemon
1 tbs. Bragg™ Liquid Aminos or 1 tsp. Real Salt™ or alternative
2 tsp. chicory root powder (sweetener)
1 heaping tbs. dried onion
2 cloves garlic
1 tbs. grated fresh ginger root
1 tbs. sesame oil
$^1/_2$ tsp. Zip (Spice Hunter) or alternative

In a blender or food processor, combine almond butter, water, lemon juice, Bragg™ Liquid Aminos, and chicory root. After this is well blended leave blender on and add the onion, garlic, ginger, oil, and Zip. Blend well and add additional water if required for thinner consistency. Can be served warmed or cool. Enjoy!

Essential Dressing

SERVES 4–6

1 cup preferred oil (Udo's, Essential Balance, olive,
    linseed, or grape seed)
1/4 cup Bragg™ Liquid Aminos or 1 tsp. Real Salt™ or
    alternative (adjust to taste)
Juice of 1 fresh lemon
1/2–1 tsp. of any seasoning you prefer, such as Italian,
    Mexican, pesto, garlic powder, onion powder, parsley,
    basil, or oregano

Combine all ingredients in a food processor and mix well or
simply place in a salad dressing jar and shake to mix well.
Chill and serve.

Basic Salad Dressing

SERVES 6

1/3 cup fresh lemon or lime juice
1 cup cold-pressed extra virgin olive oil, or any other
    preferred oil, such as Essential Balance
1/2 tsp. ground oregano
1/2 tsp. ground cumin
1/2 tsp. garlic powder
1/2 tsp. Zip, by Spice Hunter or a dash of cayenne pepper
1 tbs. Bragg™ Liquid Aminos

Put all ingredients in a blender or food processor and blend
until smooth.

Parsley Dressing

SERVES 4

1³/₄ cups water
2 celery stalks
¹/₃ cup preferred oil (virgin olive, Essential Balance,
   linseed)
1–3 cloves garlic
¹/₄ cup parsley
Optional: Salt, Bragg™ Liquid Aminos, or other
   seasonings of choice

Blend all ingredients. Use on salad and veggies.

Ginger/Almond Dressing

SERVES 4

3 spring onions (white part only)
Peeled fresh ginger (3-inch piece)
2 garlic cloves
4 tbs. almond butter
1 tsp. linseed oil
Bragg™ Liquid Aminos, to taste
1 cup water, add more if desired
Optional: 1–2 sun-dried tomatoes packed in olive oil

In a food processor, process the spring onions, ginger, and garlic
until smooth. Add almond butter, oil, and Liquid Aminos
and process until the mixture is blended. Slowly add the
water to desired consistency, and continue processing until
well blended. Serve on nonstarchy veggies, salads, and so on.

SOUPS

Creamy or Crunchy Broccoli Soup

SERVES 4–6

This high-protein soup is a must for broccoli lovers! And it takes just 15 minutes to prepare.

2 cups vegetable stock or water
3–4 cups broccoli, chopped
1 red bell pepper, chopped
2 red or yellow onions, chopped
1 avocado
1–2 stalks of celery, cut in large pieces
Bragg™ Liquid Aminos, Real Salt™ or alternative to taste
Cumin and ginger, to taste (experiment with different spices!)

In an electric skillet, warm two cups of water or stock, keeping the temperature at or below 118°F/48°C (finger test). Add the chopped broccoli and warm for five minutes.

In a blender, purée the warmed broccoli, bell pepper, onion, avocado, and celery, thinning with additional water if necessary to achieve the desired consistency. If desired, save the broccoli stalks, peeling off the tough outer skin, process them in a food processor until they are small chunks, and add to the soup just before serving to add crunch!

Serve warm, flavoring with Bragg's, fresh ginger, cumin, or any other spices you like. Add a slice of lemon on top to garnish.

Green Raw Soup

SERVES 4–6

This is a wonderfully alkalizing soup that I prefer served cold in the summer months and warmed in the winter months. It's energizing and easy to digest.

1–2 avocados
1–2 cucumbers, peeled and seeded
1 jalapeño pepper, seeded
½ yellow onion, diced
Juice of ½ lemon
1–2 cups light vegetable stock or water
3 cloves roasted garlic
1 tbs. fresh coriander
1 tbs. fresh parsley
1 carrot, finely diced

Purée all ingredients (except onions and carrot) in a food processor or blender. Use more or less water for desired consistency. Add onions and raw crunchy carrot bits at the end for a garnish. Yum!

AsparaZincado Soup

SERVES 3–5

This great soup is rich in zinc and has a rich tomato flavor—and takes only 15 minutes to prepare.

12 stalks medium asparagus (or 17 thin stalks)
5–6 large tomatoes
1 cup fresh parsley
3–5 sun-dried tomatoes (bottled in olive oil)
1–2 tsp. 3¼ pints/1.7 liters Herbes de Provence
2 tsp. Spice Hunter's Deliciously Dill or alternative
4 cloves fresh garlic
¼ cup dried onion
1 red bell pepper
1 avocado
Bragg™ Liquid Aminos to taste
2 lemons or limes, cut in thin slices

Trim and dice the tips from the asparagus and set aside for garnish. In a food processor or blender mix the asparagus and red tomatoes, parsley, dried tomatoes, spices, garlic, onion, and red bell pepper. Then blend in the avocado until soup is smooth and creamy. Warm in an electric skillet and garnish with lemon or lime slices on top. Season with Bragg's to taste or serve cold in the summertime. Sprinkle diced asparagus tips on top of the soup just before serving. Yummy!

Sautéed Edamame (Soybean) Vegetable Soup

Serves 6–8

This soup gets its wonderful silky broth from straining it after you purée it—an extra step, but well worth it. If you sauté the veggies until they are brown, you will notice a wonderful roasted flavor in this soup. The edamame (soy) beans add sweetness.

1 tbs. grape seed oil
3 large tomatoes, cored and quartered
1 large onion, sliced
1 large red pepper, cut into ½-inch strips
1 carrot, sliced
1 clove garlic, chopped
3¼ pints/1.7 liters veggie broth
3 large sprigs of parsley
¾ tsp. dried thyme
1 10-oz. pack of edamame soy beans (found in frozen
    section of your health food shop)
Real Salt™ or alternative to taste

Heat oil in bottom of soup pot, on high heat. Add the tomatoes, onion, red pepper, carrot, and garlic. Sauté until vegetables are a deep golden brown, about 15 minutes. Then take about half of those veggies out of the pan, chop fine, and set aside.

Add the veggie broth, parsley, and thyme to the remaining vegetables, heat to a boil, then cover, turn heat down, and simmer about 10 more minutes. Remove from heat and cool slightly.

Working in batches, place the soup into a blender and purée. Transfer to a strainer set over a bowl. Pour purée through the strainer and coax the liquid through the strainer into the bowl.

Discard the solids left in the strainer and return the silky broth and the remaining chopped vegetables to the soup pot. This broth should be a beautiful orange color. Add salt to taste at this point. I add a teaspoon. Add edamame (soy) beans and simmer until heated through, about five minutes longer.

Popeye Soup

SERVES 4–6

This is a wonderful alkalizing soup because of the cucumbers and greens. It is ready in just 10 minutes. Serve warm with a fresh tortilla for dipping.

1 avocado
1 cup water or yeast-free vegetable stock
2 cucumbers, unwaxed
1 cup fresh raw spinach
2 green onions
1 clove garlic
$^1/_3$ red bell pepper
Bragg™ Liquid Aminos, Real Salt™ or alternative to
    taste
Middle Eastern spices ($^1/_2$–1 tsp. garam masala, $^1/_2$–1 tsp.
    curry seasoning, and $^1/_2$ tsp. cayenne pepper)
Fresh lime juice, to taste
4 spearmint leaves to garnish

In a blender, add the avocado and half the water or stock and purée, then add the rest of the ingredients (except the spearmint leaves) one at a time, blending to desired thickness and thinning with the remaining water if desired. Add Bragg™

Aminos or salt to taste, and flavor with spices and lime juice as desired. You might add a couple of minced sun-dried tomatoes too. Experiment! Also this soup is good while on the liquid Cleanse.

Warming options: This soup can be served warm or cold. If blending in a Vita-Mix®, the longer you blend, the warmer the soup will get. If you do not have a Vita-Mix®, you can carefully warm the soup (not cook it) in an electric or stovetop skillet on low heat. Warm the soup only until you can hold your finger in it without having to pull it out. This will keep the food at about 118°F/48°C, which will keep the food raw, but warm and not cooked. Serve with spearmint leaves on top. Enjoy!

Healing Soup

SERVES 6–8

This soup is good any time, and even when you are on the Cleanse. It is soothing when you are tired or stressed, or if you have a cold or flu, and it is very antifungal.

2–3 whole cloves of garlic
1 large whole onion
3¼–5 pints/1.7–2.25 liters water
3 tbs. yeast-free instant vegetable broth
1 cucumber
1–2 carrots (optional)
2–3 tbs. fresh diced ginger
2 tbs. fresh coriander
Real Salt™ or alternative to taste
1 small head cabbage or broccoli (optional)
3 stalks celery (optional)

Crush garlic cloves and lightly steam fry. Set aside. Put whole onion in water in a deep pan, simmer until onion is transparent (approximately one hour). Add garlic and vegetable broth. Slice cucumber and any of the optional veggies you are using, and add to the soup. Simmer 10–15 minutes. Add ginger, coriander, and salt, adjusting according to taste.

Variation I: You could also bring the water to a boil, then take it off the burner and drop assorted finely chopped veggies into the water. This would just warm, but not cook the vegetables.

Variation II: You can grate, juice, or food process the ingredients into a wet paste and then add them to hot water.

Broccoli/Cauliflower Soup

SERVES 4

1 cup cucumber juice or veggie broth
$^1/_2$ cup soaked almonds
1 clove garlic, minced
1–2 cups broccoli, chopped
1–2 cups cauliflower, chopped
$^1/_4$ tsp. cumin
$^1/_4$ tsp. curry powder
1 tbs. lemon or lime juice
1 tbs. Bragg™ Liquid Aminos
$^1/_2$ tsp. Real Salt™ or alternative

In a food processor or blender, combine the almonds with the cucumber juice or broth and the garlic. Blend well. With machine still running, add the broccoli and cauliflower and blend until smooth. Finally, blend in seasonings and lemon or lime juice, Bragg™ Liquid Aminos, and salt. Add more broth or water to desired consistency.

Variation: Use an avocado instead of the almonds and use this recipe for a salad dressing.

Celery/Cauliflower Soup

SERVES 6–8

1 onion, peeled and chopped
1 tbs. oil (olive or Udo's)
1 whole head celery, trimmed and chopped (save some
**   celery leaves for garnish)**
1 head cauliflower, trimmed and chopped
1½ –3¼ pints/800 ml.–1.7 liters vegetable stock
¾ –1½ pints/400–800 ml. almond milk
Salt, pepper, and seasonings of choice, to taste

Steam fry the onion in a little water and oil in a large soup pan for about five minutes without browning. Pulse-chop the celery and cauliflower in the food processor until finely chopped.

Add the celery and cauliflower mix to the pan and warm until tender. Add the vegetable stock and almond milk and simmer for about 15–30 minutes, or you can leave this raw and not cook at all.

Purée the soup mixture in a blender or food processor until smooth texture is achieved. Season with salt and other seasonings of choice. Serve warm or cold.

Chunky Veggie Soup

SERVES 4

2¹/₂ cups fresh carrot juice
1 avocado
6–8 celery stalks
1 summer squash
2 carrots
Small bunch of arugula
Spice options: parsley, basil, or coriander

For broth, blend carrot juice, avocado, and three or four celery stalks. Grate squash, carrots, and remaining celery, adding finely chopped arugula and other fresh green spices last. Serve in bowl or cup, decorate with fresh herbs.

Broccoli Creamed Soup

SERVES 6

¹/₂ cup celery, chopped
1 chopped onion
2 tbs. oil
³/₄ pint/400 ml. vegetable broth
2 cups pure water
4 cups chopped broccoli
1 medium to large turnip, cut
1 heaped tsp. lecithin powder
¹/₂ tsp. Real Salt™ or alternative
Dash white pepper

Carefully sauté celery and onion in oil. Add broth, one cup of water, and broccoli. Cook over medium heat until broccoli is tender-crisp. Meanwhile, steam turnip until hot, but not very soft. Allow to cool slightly and purée with enough water to get a thick, smooth consistency. Add lecithin to blender and continue for a few seconds to mix. Add purée to soup and season. Cook for a few minutes to thicken.

Anticancer Soup

SERVES 2

2 tbs. caraway seed
2 broccoli stalks
2 slices each, purple and green cabbage
2 carrots
2 green onions, cut
2 cups hot water
3 tbs. fresh dill weed
1 red pepper, sliced

Soak caraway seeds in pure water 24 hours before use, then pour off liquid. Put finely cut broccoli florets and thinly sliced, peeled stems in a pot with a cover. Grate cabbage and carrots over broccoli and add caraway. Blend onions in two cups hot water and pour broth over veggies. Cover, let steam for five minutes, garnishing with dill and sliced red pepper (20 minutes to prepare).

Mock Split Pea Soup

SERVES 4

2 carrots, shaved
2 celery stalks, cut as desired
6 sprigs parsley
1 onion
4 cups water or vegetable stock
2 cups crisp steamed green beans
1¹/₂ cups crisp steamed asparagus
Dash of mace
1 bay leaf
¹/₂ tsp. Real Salt™, Bragg™ Liquid Aminos or alternative,
    to taste
¹/₂ tsp. cumin, dill, *or* Herbes de Provence (optional)

Chop all vegetable ingredients in food processor and add to
four cups of water or vegetable stock and bay leaf in a soup
pot. Lightly simmer until vegetables are just softened, about
10 minutes, remove bay leaf, then put contents into blender
and thoroughly purée until thick, creamy texture is achieved.
Add salt and seasonings. Serve warm.

Vegetable Minestrone

SERVES 4

2 carrots
2 celery stalks
1 small cabbage
1 red bell pepper
1 onion

1 courgette
1 yellow summer squash
Linseed oil, to taste
Bragg™ Liquid Aminos, to taste
Cayenne pepper, to taste

Cut vegetables as preferred. Cover carrots and celery with water or vegetable broth in soup pot. Cook gently until they just begin to "give," then add remaining vegetables. Do not over-cook. Serve hot with linseed oil, Bragg™ Liquid Aminos, and cayenne pepper to taste.

Celery Soup

SERVES 2

4–5 stalks celery (including leaves, if fresh)
3 cups pure water
2 tbs. yeast-free instant vegetable broth
Linseed oil, to taste
Bragg™ Liquid Aminos, to taste
Cayenne pepper, to taste

Cook celery until tender. Add water and broth mix. Pour all into blender. Blend 15–20 seconds. Reheat and serve. Use linseed oil, Bragg™ Liquid Aminos, and cayenne pepper to taste.

Special Carrot Soup

1 small onion, chopped
4 large carrots, sliced
1 garlic clove, minced
1 tbs. oil
$\frac{1}{4}$ tsp. turmeric
$\frac{1}{4}$ tsp. ginger
$\frac{1}{4}$ tsp. mustard seed
$\frac{1}{4}$ tsp. cumin
$\frac{1}{4}$ tsp. Real Salt™ or alternative
Pinch of ground cinnamon
Pinch of cayenne pepper
7 cups water
$\frac{1}{3}$ cup kuzu root
1 tsp. lecithin liquid or powder

In a saucepan, steam fry the onion. Add carrots, garlic, mustard seeds, spices, and salt. Cook for two to three minutes, stirring constantly. Add half a cup of water, cover, and simmer until carrots begin to soften. Let cool.

In a large saucepan bring five cups water to a near boil and reduce to medium. Stir kuzu root into one cup cool water. Slowly pour into heated water and cook until thick.

Place cooled carrot mixture in a blender and purée on low speed until smooth, adding a bit of water if needed. Add purée to thickened water and cook for five minutes, stirring as needed. Add lecithin and stir for a minute. Adjust thickness if desired.

Creamy Vegetable Soup

SERVES 8

This rich soup gets its creaminess from tofu. Be sure to blend it thoroughly (I think the blender is best) so you get a rich, even, smooth, creamy texture.

1 cup chopped onion
2 cloves garlic, minced
2 cups shredded green cabbage
3 celery stalks, chopped
$^1/_2$ lb. asparagus, cut small
2 large leeks, chopped
4 cups vegetable broth
2 tbs. chopped fresh parsley
2 tsp. dried dill
2 tsp. dried basil
1 tsp. dried oregano
Real Salt™ or alternative and pepper, to taste
1 pack soft tofu

In a skillet, steam fry the onions and garlic for a few minutes. Add cabbage, celery, and asparagus. Transfer to a large pot and add the leeks and vegetable broth. Stir in the parsley, dill, basil, oregano, salt, and pepper. Simmer just to brighten veggies. Let it cool a bit, then purée in a blender or food processor two cups at a time with some of the tofu, and return to another pot. Heat soup, not to exceed 118°F/48° C, and serve.

Gazpacho

SERVES 6 (¾-CUP SERVINGS)

4 cups tomato juice (you make)
½ cup cucumber, chopped
¼ cup green bell pepper, chopped
¼ cup celery, finely chopped
1 tbs. olive oil
½ tsp. pepper
1 tsp. basil
½ tsp. garlic, minced

Combine all ingredients. Cover and chill overnight.

Madrid Gazpacho

SERVES 6–8

3 large tomatoes
2 cucumbers
1 red bell pepper
1 small jalapeño pepper
1½ pints/800 ml. pure water
3 tbs. olive oil
2 lemons, juiced
1 tsp. cumin
2 tsp. Real Salt™ or alternative
Garlic, to taste

Blend vegetables, then add water, oil, juice, and spices. Blend again (in batches if necessary). Serve chilled, garnished with chopped tomatoes, celery, green onions, cucumbers, red peppers, and avocados.

Roasted Butternut/Celery Soup with Caramelized Onions

This soup is a satisfying soup for chilly autumn and winter days. It is also delicious made with pumpkin and makes a great breakfast, lunch, dinner, or snack.

2 butternut squash
3 celery stalks, cut in big chunks
1 onion, peeled and chopped in big chunks
1 onion, peeled and sliced into thin rings for garnish
2 tbs. olive or Udo's oil
3–4 cups veggie stock
Cinnamon and nutmeg or salt and pepper, to taste

Cut squash in half and remove seeds. Lightly oil the cut side of the squash and chunks of celery and onion. Place squash cut side down and celery and onion chunks on an oiled baking sheet and roast in a 400°F/200°C/gas 6 oven for about 45 minutes or until tender and lightly browned. Scoop out soft squash from the skins.

Purée the roasted vegetables in a blender or food processor with some of the stock. If you'd like a smoother texture, pass the soup through a strainer into a clean pot. Add the rest of the stock, season to taste, and keep warm.

To make the onion ring garnish, fry the onion in oil for 10

minutes until brown and somewhat crisp. Top soup and serve immediately.

Veggie Borscht

SERVES 8

6 cups veggie broth
1 cup carrots (shredded)
1 cup beetroots (roughly chopped)
1 cup onions (thinly sliced)
1½ cups cabbage (shredded)
1 red pepper (shredded)
Vegetized, Real Salt™ or alternative, to taste
Pepper, to taste

Combine broth, carrots, beetroots, and onions in a large saucepan. Cook gently until the vegetables are tender. Add cabbage and red pepper, salt and pepper to taste, and cook for five minutes more. The soup will have a richer flavor if cooled completely before serving time and then reheated.

Sweet Pepper Consommé

SERVES 6

3 medium red peppers
2 tomatoes
1 medium onion
¾ tsp. Real Salt™ or alternative

1 whole clove garlic
3³/₄ pints/1.7 liters boiling water

Cut the peppers in quarters and remove seeds. Quarter tomatoes
and onion. Put all ingredients in boiling water. Simmer, covered,
for one and a half hours. Strain and taste for seasoning. A
delicate and delicious broth, which may be served hot or cold.

Thick Purée of Haricot (White) Bean Soup

SERVES 8

2 lb. dried haricot (white) beans, washed and picked over
2 onions, chopped
3 large cloves garlic, minced
7 cups water
1 bay leaf
2 sprigs parsley
1 whole leaf Swiss chard, sliced crosswise
Real Salt™, alternative or Bragg™ Liquid Aminos, to taste
Black pepper, freshly ground, to taste

Soak beans for 24 hours, in three times their volume of pure
water, and drain. Steam fry onions and one clove garlic until
onions are tender. Put in large pot with seven cups water, add
drained beans, remaining garlic, bay leaf, parsley, and chard,
and bring to a boil. Reduce heat, cover, and simmer for one
hour. Add salt and pepper and continue to simmer until beans
are tender. Remove bay leaf and parsley. Purée soup in batches
in a blender. Return to pot and adjust seasonings. This can be
frozen.

Chilled Cucumber Refresher

SERVES 6

4 cups yeast-free vegetable broth
1 cup cucumber, shredded
Dill weed

Combine broth and cucumber, chill. Sprinkle each serving with dill weed.

ENTRÉES

___

Blackened Herbed Fillets

SERVES 4–6

This is a somewhat spicy coating that could be used for red snapper, trout, salmon, or even extra-firm tofu that has been sliced thin (^{1}/4 to ^{1}/3 inch/7 to 10 mm). I usually double or triple this recipe, because I like a heavy coating on the fillets.

^{1}/2 **cup grape seed oil**
3 tbs. paprika
2^1/2 tsp. dried onions
^{1}/2 **tsp. Spice Hunter's Zip or cayenne pepper (start with**
    ^{1}/4 **tsp. You can always add more.)**
1^1/2 tsp. dried thyme
1^1/2 tsp. dried oregano
1^1/2 tsp. dried basil
^{3}/4 **tsp. ground cumin**
1 tsp. Real Salt™ or alternative
^{1}/4 **cup mint leaves (fresh), minced for garnish**
Lemon and lime wedges
4 to 6 fresh fish or tofu fillets (4 to 6 oz.)

Combine all dry seasonings except mint in a shallow bowl and mix well.

Put oil into another shallow bowl and place bowls side by side.

Heat electric frying pan or large skillet-type pan on the stove to high heat.

Dip fish fillets in oil and coat well, then dip into herb mixture and coat both sides.

Cook in the hot frying pan on one side until herbs turn dark but not burned (one to three minutes), then flip over and cook on the other side.

To cook fish more thoroughly, sometimes I fry the fish first, on a medium heat, for a while until fish is cooked through, then I dip the fish in the oil and coat and fry on the outside for a crisper finish.

Sprinkle minced mint on top of fillets and garnish with lemon and lime wedges before serving.

Tofu Salad Spread

SERVES 4 (MAKES 3½ CUPS)

8 oz. *fresh* tofu, well drained
¾ cup chopped green onion
1 cup finely chopped celery
¾ cup finely chopped carrot
6 tbs. Mock Mayo (see page 349)
1 tbs. dried parsley
¼ tsp. basil
¼ tsp. sage
¼ tsp. thyme
1½ tsp. vegetable salt, Real Salt™ or alternative

$^1\!/_2$ **tsp. garlic powder**
$^1\!/_8$ **tsp. cayenne pepper**

Mix all ingredients. Serve on a bed of greens.

Tofu Patties

SERVES 6

1 carton *fresh* tofu, drained
3 tbs. onion, chopped
$^1\!/_2$ **tbs. vegetable broth mix**
1 cup courgette, grated
$^3\!/_8$ **tsp. Real Salt™ or alternative**
Egg replacer equal to 2 eggs

Slice tofu and steam for 5–10 minutes. Chop and drain well. Steam fry onions. Add vegetable broth mix and courgette and stir well. Add salt, tofu, and egg replacer and combine all ingredients. Make into patties. Put on lightly oiled baking sheets and flatten slightly. Bake gently at 350°F/180°C/gas 4. Turn patties when bottoms are barely brown. Finish baking—do not overbake.

Sprouted Bean Casserole

SERVES 6

1 large onion, chopped
1 clove garlic, finely chopped

3 cups chopped leeks
3 tbs. Bragg™ Liquid Aminos
Freshly ground pepper, to taste
1 large red or green pepper, finely chopped
1 cup pinto beans, sprouted
1 cup mung beans, sprouted
1 cup baby lima beans, sprouted

Steam fry the onion and garlic. Add leeks, Bragg™ Liquid Aminos, and pepper. Simmer 15 minutes. Add chopped pepper and simmer for five minutes. Pour over beans in casserole dish. Stir gently. Bake at 350°F, 180°C, gas 4 for 15 minutes.

Tofu Stew

SERVES 8

2 medium onions, sliced
3 cups water
1 bay leaf
3 kale leaves, torn to bite-size pieces
1½ cups string beans
2 leeks, cut to bite-size pieces
3 large onions, quartered
1 pkg. *fresh* tofu, firmness of choice

In a six-pint pan with a lid, steam fry the sliced onions. Add water, bay leaf, and kale. Cover and simmer until kale begins to soften. Remove bay leaf. Add string beans, leeks, and quartered onions and continue to simmer until beans are tender.

Meanwhile, drain and slice tofu and add to the pan to warm, or steam separately in steamer. Season if desired. Arrange tofu on top of stew to serve.

Doc Broc Casserole

SERVES 6

This is a great casserole that little kids and big kids alike will love—especially the herbed potato crisps sprinkled on top.

Florets from 2 large bunches of broccoli (save leaves and stalks, peel and clean stalks)
1 cup soft tofu
1 tsp. ground mustard seed (I use hot)
1 small bunch of fresh basil or fresh tarragon, stemmed and minced
Real Salt™ or alternative and pepper, to taste
⅔ cup olive oil
1 pack soy protein substitute that resembles ground beef
1 pack herbed potato crisps (unsalted)

Steam broccoli with a little water in a covered pan for about four to five minutes until broccoli is bright green and just crisp/tender.

Mince the broccoli leaves and peeled stalks in food processor until very fine (scrape sides down if necessary).

Add the soft tofu, mustard, basil or tarragon, salt, and pepper into the food processor with the fine broccoli mixture and process. While machine is running, slowly add the olive oil

until well emulsified and let processor run until mix is creamy
and mixed well.

In a large frying pan, heat a small amount of oil, and add the
soy substitute, crumble it up and fry it for a couple of minutes,
then add the steamed broccoli and pour the sauce (from the
food processor) over the top and stir in and mix well.

Take the bag of crisps and mash them while in the bag
with a rolling pin until they are crumbs. Then sprinkle over
the top of the broccoli mixture and serve.

Ashley's Vegetable Nori Roll-Ups

SERVES 4–6 (MAKES 2 TO 3 ROLLS)

Juice of 1 lemon
2 tbs. Bragg™ Liquid Aminos
1 tsp. extra-virgin olive oil or linseed oil
Dash of cinnamon or cayenne pepper
1 pack nori sheets
2 carrots, slivered
3 spring onions, slivered
1 avocado, slivered
1 courgette, slivered
1 cucumber, slivered
Alfalfa sprouts
Buckwheat or sunflower seed sprouts
2 cups cooked rice (basmati or brown)

In a small bowl, combine the lemon juice, Bragg™ Liquid
Aminos, oil, and cinnamon or cayenne. Place the vegetables
in a shallow pan and pour the lemon juice mixture over them.
Set aside.

Drain the vegetables thoroughly by tossing them in a colander or blotting them with paper towels.

Spread a thin layer of rice over each nori sheet, leaving about a ⅓-inch/1 cm nori border at the end. Arrange the marinated vegetables on nori sheets, top them with a lot of sprouts, and roll them up (I use a sushi mat for this). Let them sit until they will hold their form and cut into bite-size pieces with a sharp knife.

Variations: Use any vegetables and sprouts you like. Also serve with dips or sauces on top.

Stuffed Cabbage Rolls

SERVES 4

1 medium head of cabbage
1 bay leaf
1 clove garlic
1 cup onion, finely chopped
1 pkg. drained *fresh* tofu (break into fine pieces)
⅛ tsp. black pepper
½ tsp. Real Salt™, or alternative vegetized salt
1 tsp. Bragg™ Liquid Aminos
½ cup vegetable broth mix
3 cups vegetable broth, cold

Grease a shallow, four-pint stove-top casserole with a tight-fitting lid. Remove and discard wilted outer leaves from cabbage. Rinse and cut in half through core. Remove eight large leaves. Shred enough remaining cabbage to yield two cups and spread into casserole dish. Add bay leaf, garlic clove, and set casserole aside. Pour an inch of boiling water into a large pan. Add the eight leaves of cabbage and the salt. Cover and simmer for two to three minutes.

Meanwhile, steam fry chopped onion, tofu, pepper, and Bragg™ Liquid Aminos. Place ¼ cup of this mixture into the center of each of the eight cabbage leaves and roll each leaf, tucking ends in. Secure with cocktail sticks and place on shredded cabbage in the casserole dish.

Stir veggie broth mix into cold veggie broth and pour over cabbage rolls with a few grains of pepper. Cover and simmer over low heat for 30 minutes. Remove bay leaf and cocktail sticks and serve.

Curried Squash Dhal

SERVES 8

You can use any kind of squash for this wonderfully warm dish. This can be made thick as a stew and served over rice or thinned down as a soup that's great to start the day on a wintry morning!

1 medium yellow onion, quartered
½ can unsweetened coconut or almond milk
3 cloves garlic, sliced
2 serrano or Thai chili peppers, seeded and diced
1 tbs. fresh ginger root, minced
2 tsp. garam masala
1 tsp. ground cumin
½ tsp. cinnamon
1 tsp. Real Salt™ or alternative
¼ tsp. turmeric
¼ tsp. ground coriander
2 cups vegetable stock or water

1 tbs. Udo's Choice® oil or olive oil
2 cups fresh tomatoes, diced
2–4 sun-dried tomatoes, minced
4 cups butternut squash, peeled and diced
2 cups black-eyed beans or lentils, cooked
2 cups spinach or kale, chopped
1 cup green peas
3 tbs. mint, minced

Combine onion, coconut/almond milk, garlic, chili peppers, ginger root, garam masala, sun-dried tomatoes, cumin, cinnamon, salt, turmeric, coriander, and 3 tbs. stock or water in a blender. Purée mixture to a paste, scraping down the sides of the blender a few times.

Heat oil in a large saucepan, then add the spice paste and cook, stirring often, for 10 minutes. Add remaining stock, tomatoes, and butternut squash. Cook over medium heat, stirring often, until squash is just tender, about 20 minutes.

Mix in black-eyed beans/lentils, green peas, and spinach/kale. Continue to cook, stirring often, until spinach/kale is tender, about 10 more minutes. Remove from heat. Taste and adjust seasonings. Stir in the mint just before serving. Yum!

Baked Falafel Fritters

Serves 8 (makes 2½ dozen)

This recipe is wonderfully fast to whip up in your food processor. Fresh coriander and the red hot chili pepper add fun color. (The red hot chili pepper is not that hot, but remember to take the ribs and seeds out of the middle first.) I serve these warm in cabbage leaves or butter lettuce leaves, rolled up as hors d'oeuvres, *but they would also make a great side dish to a big salad or could even be thrown into a wrap or pitta sandwich! These are very hearty because they have both chickpeas and beans, making them high in calcium and protein. I use different kinds of beans to change the flavor and color of the fritters. The other plus is that they are baked instead of deep-fried like most falafel.* Bon appetit!

$^1/_4$ cup fresh parsley, coarsely chopped

$^1/_4$ cup fresh coriander, coarsely chopped

$1^1/_2$ cups tinned chickpeas, rinsed and drained (15-oz./400g tin)

8 oz./200 g. (1 cup) beans, soaked overnight (drain well and cook in boiling water for about 10 minutes, or you could use tinned in a pinch. I use black-eyed beans, kidney beans, or lima beans)

1 clove garlic, minced

1 tsp. cumin

1 tsp. turmeric

1 tsp. salt

$^1/_4$ cup red onion, chopped

1 red hot chili pepper, seeds and ribs removed, minced

1 tbs. fresh lime juice

3 tbs. flour (spelt, millet, or whole wheat)

2 heads butter lettuce or Savoy cabbage, leaves separated, tear big ones in half

6 cherry tomatoes, quartered, or 1 small tomato, finely chopped

Tahini Tofu Sauce (see below)

1 tbs. toasted or raw sesame seeds

In the bowl of a food processor, process the parsley and the coriander until fine. Add the chickpeas, beans, garlic, cumin, turmeric, salt, red onion (keep a handful of the chopped onion aside for the garnish), chili pepper, and lime juice. Pulse until the mixture forms a very thick paste that is fairly smooth (this will involve scraping the sides down and processing a few times). Add the flour and pulse to combine. Transfer the mixture to a bowl and set aside. This mixture can be made one day ahead and refrigerated in an airtight container.

On a nonstick cooking sheet drop falafel mixture 1 tbs. at a time and bake at 350°F/180°C/gas 4 for 10–12 minutes. You can also brush these with olive oil and bake until golden brown if you like.

Serve each fritter warm on a piece of lettuce or cabbage cup. Garnish with the additional onions, tomatoes, a dollop of Tahini Tofu Sauce, and a sprinkling of sesame seeds. Wrap the cabbage around the fritter and eat like a finger food *hors d'oeuvre*, or serve by a salad for a great meal!

Tahini Tofu Sauce

SERVES 2 (MAKES ⅔ CUP)

Serve this creamy, cool sauce with Baked Falafel Fritters (see above), or use as a dip for fresh veggie stix or on top of steamed veggies. Can also be used as a main spread in a wrap!

1 garlic clove, finely chopped
2 tbs. (or more) fresh lemon or lime juice
¼ cup tahini
1½ tsp. Bragg™ Liquid Aminos, or ½ to 1 tsp. salt
1 tbs. olive oil

¹/₃ **cup soft silken tofu**
Sesame seeds, raw or toasted, for garnish

In the food processor, put the garlic, lemon juice, tahini, and Bragg's™ Liquid Aminos. Process until combined. With the machine running, slowly add the olive oil through the feed tube, then add the soft tofu and pulse until smooth. Garnish with sesame seeds. This sauce can be stored in the refrigerator in an airtight container for a day or two.

To dry-toast sesame seeds, heat a heavy skillet over medium-low heat. Add the sesame seeds and shake the skillet gently to move the seeds around so that they toast evenly and do not burn. Toast the seeds until they are aromatic and barely take on color. Allow them to cool slightly before serving.

Popeye Mousse Pie

SERVES 6–8

This recipe uses psyllium seed powder to act as the binding agent that helps it set. Ask your health food shop for this item, or you can buy psyllium seed and grind it into a powder in a small coffee grinder. Makes one 9-inch/23 cm pie.

Almond Crust:
2–3 cups almonds, soaked 8–12 hours
1 tbs. Bragg™ Liquid Aminos
2–4 tbs. water or lemon juice
1 tbs. psyllium powder
1 tsp. Garlic Herb Bread Seasoning (Spice Hunter) or
    alternative

Popeye Mousse:
5 cups coarsely chopped spinach
1 stalk celery, coarsely chopped
$\frac{1}{3}$–$\frac{1}{2}$ cup lemon juice
1 tbs. Bragg™ Liquid Aminos
1 tbs. water
$\frac{1}{2}$ cup firm tofu
$\frac{1}{2}$ cup raw pine nuts
1 tbs. red onion, chopped
1 tbs. freshly chopped dill
2 tsp. psyllium powder

For the crust, process the nuts in your food processor until they are uniformly fine. Add the Bragg™ Liquid Aminos and pulse-chop. Gradually add the water or lemon juice until mixture holds together. Finally, sprinkle the psyllium and seasoning while the processor is running. Press into a 9-inch/23 cm dish and set aside.

For the mousse, put in the food processor the spinach, celery, lemon juice, Bragg™ Liquid Aminos, and water, and process until smooth. Add the tofu, pine nuts, onion, and dill. Process until the mixture is a thick purée. Gradually sprinkle in the psyllium while the machine is running. Press the pâté into the pie crust immediately. Refrigerate for at least 30 minutes or up to 24 hours before eating.

Pretty Ribbon Quiche

SERVES 6–8

This recipe is beautiful with rich colors of raw veggies. It is set by using psyllium seed powder. Use the almond crust recipe from the Popeye Mousse Pie (page 300) and build the three layers of the Ribbon Quiche in it.

Layer One:
2½ cups coarsely chopped spinach
2½ cups coarsely chopped green cabbage
½ cup fresh lemon juice
½ tsp. Real Salt™ or alternative
½ cup tofu (firm)
1 tbs. red onion, finely chopped
1 tbs. Deliciously Dill (Spice Hunter) or alternative
½ cup raw pine nuts
1½ tsp. psyllium seed powder

Layer Two:
⅓ cup tofu (firm)
4–5 carrots, chopped
1 orange bell pepper
½–1 tsp. psyllium seed powder

Layer Three:
⅓ cup tofu (firm)
2–3 medium beetroots, peeled and chopped
1 red bell pepper
1 small tomato
½–1 tsp. psyllium seed powder

Layer one: put all ingredients in food processor except psyllium seed powder. Process until well blended and somewhat smooth. With the machine still running, add the psyllium seed powder slowly. Mix well and pour this into the almond crust. Place in fridge to set while you make layer two.

Layer two: Process tofu, carrots, and orange bell pepper until smooth. With machine still running, add psyllium seed powder and continue to process until well blended. Spread evenly over layer one and place back in fridge while you make layer three.

Layer three: Follow same directions as for layers one and two and spread onto layer two. Place in fridge to finish chilling for four to six hours or overnight. Cut and serve with favorite garnish or dressing on top if desired. Ginger/Almond Dressing (see page 269) is nice.

Shelley's Super Wraps

SERVES 4

Wraps are today's answer to healthy fast food. If you stock your refrigerator with some "basics," you can make a wrap in just a few minutes. They travel well, and if you include your favorite spices, they can be "to-die-for" delicious!

1 cup Yummus Hummus (see page 363) or other hummus
1 jar non-dairy pesto or Spring's Pesto recipe (see page 349)
1 head romaine lettuce *or* any other preferred lettuce or greens (most of the time I just dig into our Rainbow Salad (see page 254) for the greens to fill in these wraps)

$^{1}/_{2}$ **cup soaked almonds**
8 sun-dried tomatoes (bottled in olive oil)
$^{1}/_{2}$ **red, yellow, orange, or green pepper, sliced thin**
2 carrots, chopped or shredded
1 cup broccoli, chopped
1 cup cauliflower, chopped
$^{1}/_{2}$ **red onion, sliced**
4–8 cloves roasted garlic
1 pack sunflower seed sprouts
8-oz./200g. pack raw pine nuts
Juice of 1 lemon or lime
Favorite spices to taste

Start with a tortilla, either one made from the Shelley's Super Tortillas recipe (opposite) or a sprouted wheat tortilla from the health food shop, or sometimes you can get a "wrap" café or restaurant to sell you its tortillas and you can keep them in the freezer or fridge. Look for hummus and non-dairy pesto from your local health food supplier.

Lay out your tortilla flat and spread with hummus or non-dairy pesto (I always use both!), or any spread you like. The Roasted Pepper Macadamia Sauce (see page 308) or even the Raw Pecan Pâté (see page 360) also work well. Then lay several leaves of romaine lettuce down the center. You could also use any other lettuce or even mixed green salad. On top of the green lettuce place any of the ingredient items. Then roll up the wrap and secure tightly in cling film (a couple of layers). Eat immediately, or at least on the same day, as the tortilla can become soggy. Enjoy!

Shelley's Super Tortillas

SERVES 6–8

4 cups flour (use any mix of flours you like, such as whole
  wheat, unbleached white, or spelt)
2 tsp. Real Salt™ or alternative
4 tsp. seasonings of your choice
2 tbs. dried onions
12 sun-dried tomatoes (packed in olive oil)
2 tsp. garlic powder
2–4 leaves fresh basil
1½ cups coconut milk or water
2 tbs. olive oil

Mix all ingredients in a food processor with a dough S-blade.
Use the pulse-chop action to prevent overheating the motor.
When the dough forms into one big ball, turn out onto a
floured flat surface and break off balls and roll them out to
about ⅛- to ¼-inch thickness. Transfer to a frying pan that
has been lightly oiled and heat on both sides until you see a
few air pockets rise. Take off the burner and let cool, then
wrap in an airtight bag and keep in the fridge or freezer. Do
not overcook, unless you want a crisped tortilla to use with
dips or soups. Or you can decrease the milk or water and add
fresh vegetable juices instead, such as spinach, parsley, or
carrot.

Nepal Vegetable Curry

SERVES 4–6

1 onion, chopped
1 bay leaf, broken
1 green chili, chopped
1 clove garlic, minced
1 inch ginger, grated
¹/₄ tsp. turmeric
Real Salt™, vegetized salt or alternative to taste
1 lb./450g carrots, cubed
¹/₂ cauliflower, broken into florets
1 cup green peas
1 tsp. each, coriander and cumin
1 cup hot water

Lightly steam fry onion. Add bay leaf, chili, garlic, ginger, turmeric, and salt. Stir in carrots and sauté lightly. Add remaining ingredients. Cook gently on medium heat until vegetables are tender.

Tofu Italian Mock Meatballs

SERVES 8–10
(MAKES 10 2-INCH OR 40 1-INCH BALLS)

This is a great transitional food when phasing off from meat. If wild rice is used, a sweet nutty flavor will result. They can be served warm for a main course, and also make a great cool snack right out of the fridge. (We always double this recipe because everyone loves them!)

1–2 cups sprouted wheat tortilla crumbs
2 stalks celery with leaves, finely chopped
1 medium red onion, finely chopped
2 cloves garlic, minced
2 lbs./900g. *firm* tofu, crumbled
1 cup vegetable stock
¼ cup whole rolled oats
3 tbs. Bragg™ Liquid Aminos
1 cup parsley
2 cups fresh basil, finely chopped
¼ tsp. black pepper, freshly ground
2 tsp. Zip (Spice Hunter) or pinch of cayenne pepper
Herbes de Provence, to taste (about 1 tsp.)
½ cup cooked brown rice
½ cup cooked wild rice
1 tbs. olive oil

Take 8–10 sprouted wheat tortillas and leave them out to dry or quick-dry them in a low-heat oven. Break them into small pieces and blend in a blender or food processor until they are finely ground into crumbs. Set aside in a bowl.

In an electric skillet, steam fry the celery, onion, and garlic and cook until softened, about six minutes. Transfer to a large bowl. Put the tofu, vegetable stock, oats, and Bragg™ Liquid Aminos in a blender and blend until smooth. Add the parsley, basil, black pepper, Zip, and Herbes de Provence and pulse until well blended. Add to the onion mixture.

Add the cooked rice to the onion mixture, along with the tortilla crumbs, and mix well. Mixture should be slightly sticky but form into balls easily. You may need to add more tortilla crumbs if mixture is too wet.

Preheat oven to 400°F/200°C/gas 6. Lightly oil a baking dish or sheet. Shape mixture into balls and roll each ball into the remaining tortilla crumbs to coat. Bake until lightly browned, 20–30 minutes. Serve with Roasted Pepper Macadamia Sauce (below) to dip the balls in. Enjoy!

Roasted Pepper Macadamia Sauce

SERVES 6–8

This is a rich, beautifully colored sauce that can be made thick for dipping with grilled tofu slices or the Tofu Italian Mock Meatballs, or it can be thinned and used as a WOW *salad dressing.*

4–5 big pieces of roasted red peppers (you can roast them or buy bottled)
6 cloves roasted garlic
3 large fresh basil leaves
¹/₂ to 1 cup olive oil
1 lb./450g. macadamia nuts (raw is best, roasted will give a different flavor)
Real Salt™ or alternative and pepper, to taste

Put roasted bell peppers, garlic, basil leaves, and a third of the oil in a food processor and process until well blended. With the machine still running, add macadamias down through the top and continue blending until well emulsified. Finally add rest of oil with machine still running and add water to thin if desired. Season to taste.

Cabbage Stuffed with Vegetables

SERVES 6

8 cabbage leaves
1 cup French-style green beans
3 tsp. dehydrated onion flakes moistened with tomato
    juice or veggie broth
2 stalks celery
$\frac{1}{2}$ cup bean sprouts
$\frac{1}{2}$ green bell pepper
1 tsp. parsley (chopped)
2 cups vegetable broth
Bragg™ Liquid Aminos, to taste
Linseed oil, to taste
Cayenne pepper, to taste

Scald cabbage leaves with boiling water and leave covered in a pot for half an hour. Chop all vegetables fine, add parsley, and mix. Spoon vegetable mixture onto each cabbage leaf. Roll tight and tuck in ends. Fasten with cocktails sticks, simmer in vegetable broth for an hour. Serve, season with Bragg™ Liquid Aminos, linseed oil, and cayenne pepper.

Cajun-Style Red Beans and Brown Rice

SERVES 8

1 lb./450g. dried pinto beans
2 cups yellow onion (chopped)
1 cup green onion (chopped)
1 cup green bell pepper (chopped)
$\frac{1}{2}$ tsp. garlic (minced)

$^1/_4$ tsp. red cayenne pepper
$^3/_4$ tsp. black pepper
$^1/_2$ tsp. Real Salt™ or alternative
$^1/_4$ tsp. oregano
$^1/_4$ tsp. garlic powder
1 oz. Bragg™ Liquid Aminos
6 oz./175g. tomato paste
$^1/_4$ tsp. thyme
1 tsp. celery flakes
6 cups cooked brown rice

Wash beans and then soak for 12 hours. Drain water. Fill a large pot with beans, add water to half an inch above beans. Add remaining ingredients except rice, cook over low heat two to two and a half hours, covered. Serve over cooked brown rice.

Beans and Rice

SERVES 4–6

2 cups dried beans such as kidney, pinto, or chickpeas, soaked and cooked according to package directions, drained
$^1/_2$ tsp. cumin
$^1/_2$ tsp. chili powder
1 tsp. Real Salt™ or alternative
1–2 cloves minced garlic, to taste
2 cups vegetable broth
$^1/_4$ cup fresh chopped parsley
$^1/_2$ onion, chopped
1 red bell pepper, chopped
2 carrots, grated

Bragg™ Liquid Aminos
2–3 tomatoes, chopped
Basmati or brown rice, cooked

Season beans with cumin, chili powder, salt, and garlic. Add enough of the vegetable broth to just cover the bean mixture. Stir in parsley, onion, pepper, and carrots and simmer until onions are soft. Add Bragg™ Liquid Aminos to taste. Mix in tomatoes just before serving with a scoop of basmati or brown rice. In a pinch, you could also use tinned beans. For a more raw and energizing dish, use sprouted beans such as chickpeas or lentils.

Stuffed Acorn Squash

SERVES 4

2 small acorn squash, halved and seeded
¼ cup water
½ cup onion, diced
½ cup carrot, diced
½ cup red bell pepper, diced
½ cup courgette, thickly sliced
½ tsp. minced garlic
Nonstick vegetable spray

Preheat over to 350°F/180°C/gas 4. Spray a large baking dish with cooking spray. Steam acorn squash halves by placing cut sides down in pan with ¼ cup water for 10–15 minutes. Lightly steam fry remaining ingredients, a few minutes only, stirring frequently. Spoon vegetables into squash halves. Bake 20–25 minutes or until squash is tender.

Autumn Curry Crêpes with Curried Veggie Filling

SERVES 6

This wonderfully colorful Thai-tasting dish can be served as an hors
d'oeuvre, *snack, or as a side dish to your main course. Serve the crêpes
fresh from the grill, as you make them, or save them in the fridge and
fill them the next morning for a really spicy warm breakfast!*

1 cup almond milk
3 tbs. unsweetened coconut milk
1½ tsp. egg replacer *or* **1½ tbs. agar agar flakes (seaweed
    gel, found in your health food shop)**
⅓ cup water
1 tbs. olive oil
½ tsp. turmeric
Dash of cinnamon
¼ tsp. curry powder
**1 cup all-purpose flour (or spelt, millet, or whole wheat
    flour)**
½ tsp. salt (optional)

In a bowl, whisk together the almond milk, coconut milk, egg
replacer or agar agar flakes, water, oil, turmeric, cinnamon,
and curry powder. Then whisk in the flour and salt until
there are no lumps left in the batter. If you are using agar
agar, then put the mixture in a food processor and process
until smooth. Put cling film over the bowl and refrigerate for
at least half an hour or up to one day.

Heat a small nonstick crêpe pan or skillet (I use my electric frying pan) over medium-low heat. If batter has begun to
separate, gently stir it to blend back in again. Once the pan is
hot, drop 2 tbs. of crêpe batter into the skillet and swirl the

pan to coat the bottom evenly with the batter. If the batter does not swirl easily, add a little water to thin it down a bit. Cook the crêpe until the top appears dry, about a minute or two. Using a spatula, gently flip the crêpe and cook until the bottom appears lightly browned and the crêpe slides easily in the pan, about a minute or two more. Transfer the crêpe to a paper towel or plate. The crêpes may be made in advance and refrigerated or frozen.

Curried Veggie Filling

SERVES 6

This filling is spicy and colorful! You can substitute veggies of your choice and it always comes out tasty! This dish is especially good in the autumn or winter because of its warming spices and grounding effect from being cooked.

$^1/_4$ **cup olive oil**

10–12 thin asparagus stalks, cut into 3-inch/7$^1/_2$ cm segments

$^1/_2$ **cup snow peas**

1 yellow onion, thinly sliced

4 cloves minced garlic

2 medium red bell peppers, seeds and ribs removed, cut into matchsticks

2 medium orange or yellow bell peppers, seeds and ribs removed, cut into matchsticks

1 tbs. fresh grated ginger

1$^1/_2$ tsp. ground cumin

1 tbs. curry powder

$^1/_2$ **tsp. cinnamon**

$^1/_2$ to 1 tsp. ground mustard seed
1 tsp. Real Salt™, Bragg™ Liquid Aminos or alternative,
   to taste
$^1/_2$ cup pine nuts
$^1/_3$ cup coconut milk (unsweetened)

Heat the olive oil in a large skillet or electric frying pan over
medium-high heat. Add the asparagus and snow peas and
cook, stirring constantly, until they barely begin to brighten
and soften. Add the onions and garlic and reduce the heat to
medium. Continue to cook until onions soften a bit. Add the
bell peppers and steam fry with a little water if necessary to
barely soften the peppers.

Add the ginger, cumin, curry powder, cinnamon, and mustard seed, and a little more olive oil, and continue to stir-mix
and cook. Add the salt, pine nuts, and coconut milk, and
cook until desired softness. I keep my veggies medium-crisp.
Serve warm with the Autumn Curry Crêpes (see page 312),
or serve over rice or any other cooked grain you prefer.

Cold Tofu Pockets

SERVES 2

1 pack firm or extra-firm *fresh* tofu
3 spring onions
$^1/_4$ cup chopped fresh coriander
$^1/_4$ red bell pepper
1 tsp. sesame seeds
1 cup Bragg™ Liquid Aminos

Soak sesame seeds overnight. Drain tofu. Cut in half on the diagonal to form two triangles, then cut a pocket in each triangle. Finely chop the spring onions, coriander, and pepper. Combine with sesame seeds. Stuff half the spring onion mixture into each piece of tofu. Poor Liquid Aminos over tofu pockets and marinate in refrigerator for 10 minutes before serving.

Vegetable Stir Fry

SERVES 4

1 tbs. oil
1 slice fresh ginger, $\frac{1}{8}$ inch pared
1 small clove garlic, crushed
$\frac{1}{2}$ cup broccoli (cut small)
$\frac{1}{2}$ cup cauliflower slices
$\frac{1}{2}$ cup red pepper strips
$\frac{1}{2}$ cup onion slices
1 cup pea pods
$\frac{1}{4}$ tsp. Real Salt™ or alternative
$\frac{1}{2}$ cup sliced celery
Bragg™ Liquid Aminos, to taste
Preferred oil, to taste
Cayenne pepper, to taste

Put oil, ginger, salt, and garlic into a wok or large skillet. Cook uncovered on heat that will not burn oil, stirring constantly for two minutes. Add broccoli, cauliflower, peppers, onion, pea pods, and celery stirring constantly for three minutes. Shut off heat, cover, let set for five minutes. Serve, flavor with Bragg™ Liquid Aminos, preferred oil, and cayenne, to taste. If broccoli is not cooked sufficiently for your taste, try slightly steaming before adding to the stir fry.

20-Minute Alkaline Stir Fry

SERVES 4

1 pack buckwheat soba noodles
$\frac{1}{2}$ package extra-firm tofu, cubed
Bragg™ Liquid Aminos, to taste
Vegetable broth, as needed
1 red bell pepper, chopped
1 onion, chopped
1 head broccoli, cut into florets, and/or
1 bunch asparagus, cut into 1-inch lengths
Olive oil
Raw, unhulled sesame seeds
Garlic
Stir fry spice combination and/or ginger

Break noodles into quarters, prepare according to package directions, and drain. While the noodles are cooking, sauté tofu, in a large pan, in a little Bragg's and vegetable broth for about five minutes. Remove from pan and set aside. Sauté veggies for about five minutes in more Bragg's and broth. Add tofu and noodles. Sprinkle with olive oil, sesame seeds, garlic, and spices and stir gently.

Vegetable Steam Fry

Serves 4

1–2 tsp. fresh grated ginger (use a hand grater)
2–3 cloves garlic, crushed
$\frac{1}{2}$ cup broccoli (cut small)
$\frac{1}{2}$ cup cauliflower slices
$\frac{1}{2}$ cup red pepper strips
$\frac{1}{2}$ cup onion slices
$\frac{1}{2}$ cup yellow squash
1 cup pea pods
(Other veggies as desired, cut julienne)
1 cup fried tofu (or use marinated tofu from the health
    food shop)
$\frac{1}{4}$ tsp. Real Salt™ or alternative

Steam-fry Sauce:
$\frac{1}{3}$ cup water or veggie stock
1 tsp. Stir-Fry Ginger Spice (Spice Hunter) or alternative
Juice of $\frac{1}{2}$ lemon or lime
Bragg™ Liquid Aminos, to taste

Heat electric frying pan. Add small amount of water and steam
fry the ginger and garlic for a couple of minutes. Add veggies,
tofu, and salt and steam fry until veggies are very bright and
slightly softened. Pour the steam fry sauce mixture over the top
and steam for one or two more minutes. Serve immediately!

Variation: Sometimes I add pine nuts or pecans to enrich
this dish.

Maren's Tortilla Pockets

SERVES 2 TO 4

This makes sealed-pocket sandwiches (up to 4 sections depending on your sandwich maker), and they are delicious! Serve with soup or salad. These little sealed sandwiches are a great way to hide the insides, too! Children especially love them. They would make a great appetizer, too!

2 large sprouted wheat tortillas or other tortillas

Filling of your choice, such as:
$\frac{1}{2}$ **cup hummus,** $\frac{1}{2}$ **cup rice,** $\frac{1}{4}$ **cup salsa**
$\frac{1}{4}$ **cup pesto,** $\frac{3}{4}$ **cup steamed veggies, 1 tsp. almond butter**
$\frac{1}{4}$ **cup salsa,** $\frac{1}{2}$ **cup black beans,** $\frac{1}{2}$ **cup cooked millet or buckwheat**
$\frac{1}{2}$ **cup avocado,** $\frac{1}{3}$ **cup salsa,** $\frac{1}{2}$ **cup grilled tofu cubes**
$\frac{3}{4}$ **cup grated veggies,** $\frac{1}{4}$ **cup sun-dried tomatoes,** $\frac{1}{4}$ **cup soaked almonds, chopped**
1 cup stir-fried veggies
1 cup Casserole de Cauliflower (see page 345)

Preferred oil
Real Salt™, Bragg™ Liquid Aminos or alternative
Spices (optional), such as Mexican, Italian, barbecue

Set up a toasted sandwich maker. Make a cardboard pattern the size of your sandwich maker. Set heat on appliance to medium to medium high and leave open. Stack your tortillas. Place the pattern in the middle and with a sharp knife cut out tortillas. (Use extra pieces to make chips by baking on a baking sheet at 325°F/170°C/gas 3 for 10 minutes.) Place one shaped tortilla on the bottom half of the appliance. Place fillings into center of each section of the press with a spoon, keeping it away from the edges. Place the second tortilla on top and close and lock lid for about three minutes, or until lightly golden.

Serve with soup or salad or use as a healthy snack. You can make these ahead of time, and they travel well. Experiment and come up with your own favorite combinations for fillings!

Sunrise Asian Salad

SERVES 4

This is a hearty, chewy salad with the addition of beans and wild rice.

$^1/_2$ **cup adzuki beans**
$^1/_2$ **cup black beans**
$^1/_2$ **cup black-eyed peas**
$^1/_2$ **cup brown rice**
$^1/_2$ **cup wild rice**
$^1/_3$ **cup fresh lemon juice**
1 Tbs. Bragg™ Liquid Aminos or 1 tsp. Real Salt™ or alternative
1 tsp. curry powder
1 clove fresh garlic, minced
1 tsp. Zip (Spice Hunter) or 2 tsp. black pepper
1-inch cube of fresh ginger, grated
$^2/_3$ **cup olive oil**
$^1/_2$ **medium red onion, sliced thin**
1 carrot, julienned (use mandolin)
$^1/_2$ **cup snow peas, trimmed and sliced**
1 cup bean sprouts or 1 cup sprouts, any kind

Place beans and rice into large bowl.

In small food processor, place lemon juice, Bragg™ Liquid Aminos, garlic, curry powder, ginger, and pepper. Gradually add olive oil and process until well emulsified.

Pour dressing over bean/rice mixture. Add remaining vegetables. Toss well and then chill for two hours before serving.

Tofu Spinach Quiche

SERVES 6–8

Pastry:
4 cups whole wheat flour
6 tbs. preferred oil
Pinch of Real Salt™ or alternative
A little cold water

Filling:
2 onions (diced)
³/₄ cup vegetable oil
2 tbs. parsley (chopped)
2 tbs. dill weed
2 cups spinach (chopped and cooked)
2 10-oz./250g. packs frozen spinach
Real Salt™, to taste
2 cups tofu
¹/₄ cup soy milk (if necessary)

Pastry:
Combine the ingredients and knead the dough into a cohesive ball.

Roll out the pastry on a floured board and press it into an oiled pie dish.

Filling:

Sauté the onions in oil until they are transparent.

Add the parsley, dill weed, spinach (defrosted), and salt and mix them in well.

Blend the tofu in a processor with the soy milk if it is difficult to blend on its own. (You might also put the parsley into the blender to chop it up more easily.)

Pour this over the vegetable mixture and mix it thoroughly.

Place the filling in the pastry case and bake it at 375°F/190°C/gas 5 for about 30 minutes.

Nutty Mock Meat Loaf

Serves 6

1 cup almonds, raw
²/₃ cup sunflower seeds, raw
¹/₂ cup brazil nuts, raw
¹/₄ cup linseed, ground
2 small onions, diced
¹/₂ cup parsley, fresh
¹/₂ tsp. Real Salt™ or alternative
¹/₂ tsp. sweet basil
¹/₂ tsp. sage
¹/₃ tsp. thyme
¹/₂ cup water

Sauce:
¹/₂ cup almonds, ground
2 cups water
1 tsp. seasoning, your choice
2 tbs. arrowroot flour

Dash of cayenne pepper
2 tbs. olive oil
¼ tsp. Real Salt™ or alternative

Preheat oven to 350°F/180°C/gas 4. Grind nuts and seeds in a processor, blender, or grinder. Combine remaining dry ingredients, mix well. Add the water and mix again. Place in a well-oiled loaf tin and bake for 25 minutes. Slice and serve with sauce.

Combine all sauce ingredients, bring to a low boil, and stir constantly. Turn heat down and simmer on low heat until thick. Pour over top of nut loaf.

Serve with tossed salad or steamed vegetables. This is a good snack and freezes well.

Alexandra's Favorite Pasta

SERVES 6

2 tsp. olive oil
28-oz./800g. tin plum tomatoes, undrained
2 cloves garlic, minced
16 oz. spaghetti or fettuccine, uncooked
8 oz./225g. almond or non-dairy cheese
⅛ tsp. red pepper flakes

Cube tomatoes, heat with juice over medium heat with garlic and olive oil for 20 minutes.

Meanwhile, cook pasta, drain, and place in serving bowl. Add tomatoes, cheese, and red pepper flakes, toss. Cover bowl for five minutes to allow cheese to melt. Toss again before serving.

Pasta with Creamy Pesto Sauce

SERVES 4–6

This is a colorful, wonderful way to veggie up your pasta. Even better, you can just eat the veggies and sauce without the pasta. I serve the sauce raw and the dish cold in the summer, and warm it all up in the winter.

3 cups shaved small yellow squash (courgette would work too)

2 red bell peppers, cleaned and sliced into long thin strips

1 red onion, sliced in thin rings (marinated in Bragg™ Liquid Aminos and/or some Real Salt™ or alternative)

1 cup chopped tomato

$^1/_2$ cup sun-dried tomatoes, chopped fine in food processor

$^1/_2$ cup fresh basil, chopped (or 1 tbs. dried basil)

2 tbs. fresh oregano, chopped (or 1 tsp. dried oregano)

1 tbs. fresh rosemary, destemmed and chopped (or $^1/_2$ tsp. dried rosemary)

$^1/_2$ tsp. Spice Hunter's Zip seasoning or alternative

1 tsp. minced garlic (or $^1/_2$ tsp. garlic powder)

1 tsp. minced ginger (or $^1/_4$–$^1/_2$ tsp. powdered ginger)

$^1/_3$ cup fresh lemon juice

1 lb./450g. vegetable pasta made without eggs

1 recipe Spring's Pesto (see page 349)

1 can coconut milk

Real Salt™ or alternative to taste

Soy parmesan (optional)

With a vegetable peeler, shave yellow squash (or courgette) lengthwise into long, thin strips, cutting any remaining parts with seeds into long strips with a knife, *or* put squash through the Saladacco machine that cuts it into angel hair pasta. In a bowl, combine with all ingredients down to the lemon juice, mix well, and set aside. Prepare pasta according to package directions. Drain and toss with a small amount of olive oil to prevent sticking. Keep warm. Put pesto and coconut milk into food processor and process until creamy. While processor is still running, add water to desired consistency. Add salt to taste. To serve, arrange pasta on a plate, top with veggie mix, then creamy pesto sauce. Sprinkle with a soy parmesan if desired.

Almond/Carrot/Ginger-Stuffed Courgettes

SERVES 4

1 large onion, peeled and chopped
2 tbs. olive oil
4 medium courgettes
1 clove garlic, minced
4 medium carrots, scraped and finely diced
1 tsp. grated fresh ginger
$2/3$ cup soaked almonds, chopped, or raw unsoaked
    macadamia nuts, chopped
Real Salt™ or alternative to taste
Zip (Spice Hunter) or pepper, to taste

Preheat oven to 375°F/190°C/gas 5.

Sauté onion in the oil in a medium saucepan for five minutes. Halve the courgettes lengthwise and scoop out the soft centers to make good cavities for stuffing.

Chop the scooped-out centers of the courgettes and add it to the onion along with the garlic, carrots, and ginger. Cover and sauté gently for about 10 minutes, until the veggies are slightly soft.

Remove from heat and add the chopped almonds or macadamias and seasonings to taste.

Place the courgette skins in an oiled shallow casserole and fill them with the carrot mixture. Cover and bake for about 30–40 minutes. Serve immediately.

Edamame Soy Patties

SERVES 4

These are a hardy little veggie burger that can be pan fried if you're in a hurry, or dehydrated for 4–6 hours to make a more crusty outside. They are made with a base of edamame (soy beans from pods) that you can find in your health food shop in the frozen section.

3 tbs. linseeds, ground into powder (using blender or
    coffee grinder)
6 tbs. water
1 carrot, grated fine or processed to fine pulp
2 cloves garlic
1 tsp. dried onion
2 sun-dried tomatoes (packed in olive oil)
$^1/_2$ cup parsley
$^1/_2$ tsp. dried mustard
$^1/_2$ tsp. turmeric
1 tsp. Mexican seasoning
$^1/_3$ tsp. Deliciously Dill (Spice Hunter) or alternative

1 tsp. Real Salt™ or alternative
10-oz./275g. pack vegetable soybeans

Grind linseeds to a powder and put in a bowl. Add the water and stir to mix well. Set aside to gel.

Put carrot, garlic, dried onion, sun-dried tomatoes, linseed mixture, and all spices into a food processor and process to desired consistency. (I do this quite smooth.) Then add the edamame beans and process until well mixed. You can make them coarse and more chunky with the beans showing or smoother and more mixed if you like. Also you can use more edamame beans if the mixture seems to be too moist. They should be able to form into a patty easily. Then make into small patties and put into a dehydrator for four to six hours, or use a little grape seed oil and dip the patties in some sprouted wheat tortilla crumbs and fry on both sides. Serve with the Rich Raw Tomato Sauce (see page 345) on top for a real treat!

Optional: You can use all sorts of different spices for these. Also you could add some chickpeas or other beans of choice to stretch the recipe or add more bean flavors. Experiment! Enjoy!

Hearty Harvest Casserole

SERVES 12

2 large onions, cut and separated into rings ¾ inch thick
1 each medium green and red pepper, cut into 1-inch strips
1 cup sprouted barley, partially cooked (save 1 cup water)
1 cup barley water (saved above)
4 tbs. vegetable broth mix

3 medium carrots, cut into chunks
2 large tomatoes, peeled and quartered
2 medium courgettes, cut into 1½-inch chunks
1 lb./450g.green string beans, snapped in half
½ head cauliflower florets
2 cloves garlic, crushed
1 tbs. Real Salt™ or alternative
¼ tsp. black pepper
1 tsp. paprika
¼ cup parsley, chopped

Steam fry onion and bell peppers. Combine all ingredients in a casserole dish. Cover. Bake at 350°F/180°C/gas 4 for an hour. Barley should be tender.

Spiced Winter Squash

SERVES 2

2 cups butternut squash, grated
2 tbs. olive oil
1 cup grated acorn or banana squash
2 tsp. garam masala or curry powder
Pinch of cinnamon
1 tbs. lemon or lime juice or both
1 tbs. Bragg™ Liquid Aminos
2 tbs. minced onion

Combine the squash, oil, spices, juice, Bragg™ Liquid Aminos, and onion. Mix well and toss. Then turn into a frying pan and gently warm this dish right before serving.

Fresh Spinach/Courgette Bake

SERVES 6

2½ tbs. oil
2 garlic cloves, 1 chopped
20 oz./550g. fresh spinach
¾ cup chopped onion
1 tbs. basil leaves
2 cups diced courgette
¾ tsp. vegetized or Real Salt™ or alternative
⅛ tsp. pepper
Egg replacer equal to 6 eggs
¼ cup sprouted wheat bread crumbs

In large, warm saucepan put 1 tbs. oil, the whole garlic clove, and washed spinach. Cover and cook for four minutes, wilting the spinach leaves. Remove garlic and drain spinach. Heat oven to 350°F/180°C/gas 4. Reheat pot, add 1½ tbs. oil, chopped garlic, onion, basil, and courgette. Cook until onion is soft. Mix in spinach and seasoning. Oil a baking dish. Spread spinach-courgette mixture over bottom of dish. Pour egg substitute over vegetables, tilting to distribute the mixture evenly. Sprinkle with crumbs. Bake 10–14 minutes or until egg substitute is set.

Zippy Breakfast (our favorite!)

SERVES 1

This would make a delicious meal any time, but we love it to start our day.

1–2 cups cooked rice or grain of your choice (I use
basmati, brown, or wild rice, millet, quinoa, or
buckwheat)
1 avocado, sliced
1 firm tomato, chopped
Juice of 1 lemon or lime (or both)
1–2 tsp. oil (linseed, Udo's, olive)
1–2 tsp. Bragg™ Liquid Aminos
Zip (Spice Hunter) or alternative to taste

Start with the warm rice or grain in a bowl. Slice the avocado
and tomato on top. Then drizzle the oil, Bragg™ Liquid
Aminos, and lemon juice over the top. Sprinkle with Zip to
taste. For other grain choices, try buckwheat groats or spelt.

Variation: Sometimes I throw in some chopped red bell
pepper, sunflower seed sprouts, and soaked almonds over the
top for extra crunch! Enjoy!

Green Chili Tofu Pitta

Serves 6

*These are great little Mexican triangles stuffed with a fresh
tofu/coriander filling. Great for a snack or appetizer or as a main
course beside a big salad.*

1 pack pitta bread or tortillas
3 cloves garlic, minced
1 small tin green chilis (chopped)
1 pack extra-firm tofu
1 tsp. Mexican seasoning (Spice Hunter) or alternative
2 tsp. dried onion, *or* ¼ cup minced fresh onion

$^1/_4$ **cup soy parmesan cheese substitute**
1 tbs. fresh coriander
$^1/_2$ **tsp. Real Salt™ or alternative**
1 jar or tin enchilada sauce
Avocado slices for garnish
3–4 sun-dried tomatoes for garnish

Take the pitta bread and cut it like a pie into eight triangular pieces and open each one up so you can put the filling in. In a food processor mince the garlic, then add all other ingredients except the tofu, the enchilada sauce, and the garnishes, and process until finely chopped. Then put the grater attachment on the processor and grate the tofu into the mix. Process for a few seconds more to mix.

Spoon the filling into the pitta triangles and place into a pie pan. Spoon enchilada sauce over each pitta and inside over the filling mixture. Bake at 350°F/180°C/gas 4 for 10–15 minutes. Cut sun-dried tomatoes and avocado slices to put on top for a garnish just before serving warm.

SIDE DISHES

Mexicali Rice

Serves 6–8

This is a salsa rice that is a great complement to a Mexican dinner. Serve with Great Olé Guacamole (see page 358) and Green Chili Tofu Pitta hors d'oeuvres *(see page 329).*

3 tbs. olive oil
1 onion, chopped
1 clove garlic, minced
$^1/_2$ cup celery, diced
$^1/_4$ cup bell pepper, diced
2 large tomatoes, coarsely chopped
2 serrano chilis, seeds and stems removed, chopped
2 tbs. fresh coriander, chopped
1 tsp. fresh lime juice
$^1/_2$ tsp. oregano
1 tsp. Real Salt™ or alternative
3–4 cups cooked rice

Put half the olive oil in a skillet or frying pan and sauté the onion and garlic until onion softens. Then add the rest of the ingredients (I chop them all up in a food processor) except the rice and olive oil and steam fry until veggies are bright and still somewhat crisp. Add three to four cups cooked rice and the rest of the olive oil. Mix well and serve warm.

Millet Yam Hash Browns

SERVES 4

Millet is a good source of iron, lecithin, and choline, and yams are high in vitamin E. This is a nice recipe to help "wean off" starchy deep-fried hash browns. I doubled this recipe and it was gone in two days!

2½ cups water
½ tsp. Real Salt™ or alternative
1 cup millet
1 yam (carrot or sweet potato works too), peeled and
    processed to semifine/coarse chunks
1 tsp. dried onions
½ tsp. Deliciously Dill (Spice Hunter) or alternative
½ tsp. Garlic Herb Bread Seasoning or Vegetable Rub
    (Spice Hunter) or alternative
½ tsp. dried garlic powder (could use fresh or roasted)
Grape seed oil for brushing
Minced fresh coriander for garnish

Preheat oven to 400°F/200°C/gas 6. Bring the water and salt to a boil in a medium saucepan.

Add the millet, lower heat, cover, and simmer for 15 minutes.

Open lid and place the processed yam on top of the millet, return the lid, and continue to let simmer for 10 more minutes. Transfer to a large bowl.

Add all remaining spices and toss. Add salt to taste, but don't douse the natural sweetness of the yams.

The mixture should be sticky and stiff enough to hold shape when formed into oval patties.

Using a ¼-cup measuring cup, scoop out the batter and form oval patties about ⅓–½ inch/7–10mm thick.

Place on an oiled baking sheet and bake for 20 minutes or until brown.

Sprinkle the minced fresh coriander with some Ginger-Almond Paste Topping on top (see page 351).

Enjoy. These are incredible right out of the oven. Later we spread hummus on top and have them as a snack.

Wild Yam Soba Noodles with Kale and Spicy Pine Nuts

Serves 4

This is a nice, filling side dish that warms the body.

1 bunch kale or fresh spinach
2 tbs. grape seed oil
3 cloves garlic, minced
⅓ cup veggie broth
2 tbs. Bragg™ Liquid Aminos
Juice of ½ lemon
1 pack Wild Yam Soba Noodles
½ cup pine nuts, spiced (see Spicy Pecan Croutons recipe on page 377)
2 tbs. crumbled nori (optional)

Slice the kale (or spinach) stems thinly and chop the leaves coarsely or put in food processor and pulse-chop until coarse.

Heat oil in a large frying pan on a medium heat. Sauté the garlic and kale stems for a few minutes, then add the leaves, veggie broth, and Bragg's. Continue to sauté until kale/spinach is tender and bright, about eight minutes. Add lemon juice and mix well. Take out of the frying pan and set aside. Note: At this point you can put the kale/spinach in a food processor and process to a finer smooth pâté if you prefer.

In the same frying pan, prepare pine nuts according to Spicy Pecan Croutons recipe (see page 377) and set aside for garnish.

In a large pot of boiling water, cook the soba noodles until tender, then rinse and drain well.

Put the cooked noodles into the frying pan and add the kale mixture on top. Mix in well with noodles and sprinkle the spicy nuts and the crumbled nori on top.

Millet/Buckwheat Pancakes

SERVES 6

These are incredibly hearty dense pancakes made of ground raw hulled buckwheat and millet. They have an "eggy" consistency and are made from a thick batter. Buckwheat has a binding quality and is gluten-free. It is a seed, not a grain, and thus it is more alkalizing. My family likes them served fresh from the oven, on the side of a big salad, or just as a snack for breakfast, lunch, or dinner. They would also travel well. I use a blender to grind the buckwheat and millet into flour. Sometimes it's best to do a cup at a time.

Experiment with this basic recipe and change the vegetables and seasonings to come up with several different versions (suggestions below).

3 tbs. grape seed or olive oil for frying
1 cup millet, ground to flour
1½–2 cups raw hulled buckwheat, ground to flour
1 small yam, peeled and shredded
1 onion, sliced or chopped
1 tbs. dried parsley (or ½ cup fresh chopped parsley)
2 cloves garlic
2 cups water, Rice Dream, almond milk, or soy milk
¼ tsp. cinnamon (I use more)
¼ tsp. nutmeg (I use more)
1 tsp. Real Salt™ or alternative
1 tsp. Vegetable Rub (Spice Hunter) or alternative
1 tsp. Mexican Seasoning (Spice Hunter) or alternative

Grind millet and buckwheat to flour in a blender. Set aside in a large bowl and mix well together.

In the blender place the shredded yam, chopped onion, parsley, spices, salt, and water/milk and blend until smooth. Pour into the dry flour mixture and mix well. Batter should be quite thick and stiff.

In a frying pan, heat the oil over medium heat and spoon the batter (about 3–4 tbs. per pancake) into the oiled pan. Cook until golden brown on one side and flip over to the other side and cook until done. You may need to add more oil as you continue to cook the rest of the pancakes. Serve hot, or serve cold as a snack.

Another option is to bake square pancakes or oven cakes (my family likes these best):

Get a 9-by-13-inch/23-by-32-cm glass dish and oil it well. Pour entire batter mixture into the glass dish. Bake in an oven preheated to 375°F/190°C/gas 5 for 30 minutes covered, then uncover and let cook 20–30 minutes more to slightly brown the top. Let cool and cut into squares. Can be served with any sauce, pesto, or dressing you like on top. The AvoRado AvoCado Topping (see page 350) works great!

Substitutions suggestions:

Try adding sesame seeds, sprouted sunflower seeds, or pine nuts for a nuttier texture.

Instead of yam, substitute a sweet potato, beetroot, or carrots.

Instead of parsley, substitute kale, spinach, or grated courgette.

Kale with Egyptian Garlic Sauce

SERVES 4

Also makes two or three side-dish servings. Kale is a tasty green and deserves to be better known. Its ruffled leaves cook to a deep green and are an appealing accompaniment for rice or other whole grains. Kale is delicious with an Egyptian sauce of sautéed garlic and ground coriander. It's not exactly a sauce, but a seasoning mixture—a way to add a quick burst of flavor to a cooked vegetable. Also good with okra or baked aubergine. It can also be mixed with brown rice and courgette.

1 lb./450g. kale
2 tsp. ground coriander

4 medium garlic cloves, minced
Real Salt™ or alternative to taste
Cayenne pepper, to taste

Rinse kale and remove stems, including the tough part of stem in the leaf. Pile leaves and cut into manageable size. Steam kale until tender-crisp, and transfer to a bowl. Steam fry garlic, about 1 minute. Add coriander, salt, and cayenne and stir over low heat for 15 seconds to blend. Immediately toss with kale, in pan or in a bowl. Taste and adjust seasoning. Serve hot.

Shelley Beans

SERVES 4

In Indiana and Ohio, Shelley beans are known as cranberry beans. This bean with its red-striped and cream color is a much sweeter bean and more delicate than pinto beans. When cooked, they lose their marking and become solid in color. This rich recipe includes macadamias, ginger, and lime, which make the beans especially creamy and tasty.

2 cups cooked cranberry beans
15–20 raw macadamia nuts
2 cloves garlic
$^1/_3$–$^1/_2$ inch piece ginger, grated on fine cheese grater
1 cup water
1 tsp. grape seed oil
1 small jalapeño chili, seeded and minced
1 tsp. ground coriander
$^1/_2$ tsp. cumin

Lime juice, to taste
Sections from 1 lime (skins included), minced in
  processor
Chopped fresh basil or coriander for garnish
1 tsp. Real Salt™ or alternative

Soak beans overnight and cook in water and 1 tsp. salt until
done. Set aside.

Combine macadamias, garlic, and ginger in a food processor. While processing add $\frac{1}{2}$ cup of water and blend to a
thick milk sauce.

In a frying pan, heat oil over medium heat. Add the jalapeño
and cook for a few minutes, then add the ground coriander and
cumin. Continue cooking for another minute. Add cooked
beans and continue to stir, warming the beans up for another
few minutes.

Stir in the nut milk/ginger mixture and add more of
the water if desired for a thinner consistency. Just before
serving, add the lime juice and minced lime sections. Stir
and mix well. Serve with the chopped basil or coriander on
top. Also add Real Salt™ to taste. Grapefruit sections make
a nice garnish.

Courgette Italian Style

SERVES 8

8–10 medium courgettes
2 cloves garlic, minced
$\frac{2}{3}$ cup onion, coarsely chopped
1$\frac{1}{2}$ cups tomatoes
1 tsp. Real Salt™ or alternative

$^1\!/_8$ **tsp. pepper**
3 tbs. olive oil

Wash, trim ends, and slice courgettes. In a saucepan, steam fry garlic, onion, and sliced courgettes over low heat, 10 minutes, turning and moving mixture occasionally.

Remove courgette mixture from heat and sieve in tomatoes with pepper and salt. Blend lightly and thoroughly.

Turn mixture into a casserole dish, cover, and simmer 30 minutes.

Add the olive oil to the dish before serving.

Okra and Tomatoes Creole

SERVES 6–8

4 cups sliced okra
1 cup chopped onion
$^1\!/_3$ **cup chopped green pepper**
2 cups chopped tomatoes
$^1\!/_2$ **tsp. Real Salt™ or alternative**
$^1\!/_8$ **tsp. black pepper**
$^1\!/_8$ **tsp. curry powder**
1 tsp. powdered lecithin
$^1\!/_8$ **tsp. thyme**

Wash okra, cut off stem ends, slice, and set aside. Chop onion and green pepper, steam fry in a large skillet to transparent stage. Add okra and tomatoes. Stir in mixture of salt, pepper, curry powder, lecithin, and thyme. Simmer, covered, 30–40 minutes, or until okra is tender.

Steam-Fried Sprouts

SERVES 4

2 tbs. Bragg™ Liquid Aminos
¹/₂ cup finely chopped onion
¹/₂ cup finely chopped green or red pepper
2 cups fresh bean sprouts, any kind

Steam fry the onion and pepper in a skillet, with Liquid Aminos. Add sprouts and steam gently for 30 seconds. Serve immediately.

Refried Beans

SERVES 6 (MAKES 3 CUPS)

¹/₂ cup onion, chopped
1 tsp. minced garlic
3 cups cooked pinto beans
Garlic powder, to taste
Cayenne pepper, to taste
Black pepper, to taste
Real Salt™ or alternative to taste

Steam fry onions and garlic. Purée pinto beans in food processor or blender. Pour puréed beans into the skillet. Stir beans constantly on low to medium heat until thickened; season while cooking. Serve hot with vegetables.

Spiced Green Beans

SERVES 6

1 lb./450g. green string beans
$\frac{1}{2}$ cup boiling water
$\frac{1}{2}$ tsp. Real Salt™ or alternative
1 cup thinly sliced onion
$\frac{1}{4}$ tsp. black pepper
$\frac{1}{4}$ tsp. nutmeg
3 tsp. linseed oil
1 tbs. parsley

Wash green string beans, break off ends, then cut lengthwise into fine strips. Carefully place beans and salt into saucepan of boiling water. Cook, loosely covered, until tender-crisp. Meanwhile, in skillet steam fry the onion. Drain beans and add to skillet with a mixture of salt, pepper, and nutmeg. Sauté for five minutes. Add linseed oil and parsley. Toss well and serve.

Casserole de Cauliflower

SERVES 4–6

Takes 20 minutes to prepare. This dish is a lot like couscous in texture and makes a great breakfast, lunch, or dinner side dish.

2 tsp. oil (olive, linseed, or Udo's Choice®)
2–4 tsp. cumin
$\frac{1}{2}$ tsp. turmeric
$\frac{1}{2}$ yellow or red onion, finely minced

1 cup water
Florets from 1 very large or 2 small cauliflowers
7–8 sun-dried tomatoes packed in olive oil)
1 red bell pepper, finely chopped
4 tbs. fresh parsley, minced
$\frac{1}{2}$ cup raw pine nuts
Bragg™ Liquid Aminos, to taste
Lemon or lime juice, to taste
2 cloves garlic, minced

In a skillet, warm the oil, cumin, and turmeric.

Keeping the temperature on warm or low, add the onion and allow the flavors to blend for three to four minutes, then add the water and warm.

In a food processor fitted with an S-blade, process the cauliflower into very small pieces (like couscous). Also process the sun-dried tomatoes into fine small pieces.

Add the cauliflower to the skillet and gradually warm, adding the parsley, garlic, sun-dried tomatoes, red pepper and pine nuts. Season with Bragg™ Liquid Aminos and lemon or lime juice to taste. Enjoy!

Ginger Beans and Carrots

SERVES 4

1 lb./450g. string beans
4 tbs. oil
1 tsp. mustard seeds
$\frac{1}{2}$–$\frac{3}{4}$ cup chopped onions
4 carrots, thinly sliced
$\frac{1}{4}$ tsp. ground ginger

1 tsp. Real Salt™ or alternative
2 tbs. fresh lemon juice

Wash and snap beans. Carefully heat the oil in a skillet. Add mustard seeds and sauté for 30–40 seconds (seeds will pop). Stir in onion, carrots, and beans. Cook, stirring, for five minutes. Stir in salt and ginger, lower heat. Cook for 10 minutes. Stir in lemon juice just before serving.

Avocado/Tomato Snack

SERVES 2–3

2 avocados
1 small aubergine, diced
1 tsp. curry powder
2 tbs. lemon juice
2 seeded green chili peppers
Real Salt™ or alternative and seasoning, to taste
2 or 3 tomatoes, thickly sliced

Blend all ingredients except tomatoes in blender until smooth. Spoon onto warmed tomato slices.

SAUCES

Lime Ginger Sauce

SERVES 4–6

This makes a wonderful sauce, dressing, or marinade.

¼ cup lime juice
¼ cup oil (linseed, olive, or Udo's Choice®)
1 tbs. Bragg™ Liquid Aminos
¼ cup water
1 tbs. fresh mint
1 tbs. fresh coriander
1 tsp. minced ginger root
¼ tsp. dried red chili pepper
2–3 tsp. fresh arrowroot or carrot juice
1 tsp. Real Salt™ or alternative to taste
Dash of Zip (Spice Hunter) or alternative

In a processor or blender combine all ingredients and blend
well.

Rich Raw Tomato Sauce

SERVES 4–6

This is a wonderfully fresh-tasting raw tomato sauce that goes great over pasta. I use it over raw yellow crookneck squash angel hair, which I make with a gadget called the Saladacco. It also is a wonderful dipping sauce, and can be served cold or warmed but not cooked.

3–5 sun-dried tomatoes (packed in olive oil)
4 fresh, firm tomatoes, chopped
¹/₂ cup fresh basil, chopped
1 tsp. dried onion
1 tsp. roasted garlic
1 tsp. Real Salt™ or alternative

Put all ingredients in a food processor and blend to desired consistency. Store in airtight container in refrigerator for up to three days.

Variation: Instead of the fresh basil and roasted garlic, just use ¹/₄–¹/₃ cup of non-dairy pesto. This is also great on wraps. It is available from some large health food suppliers.

Spring's Pesto

SERVES 4

6 cloves garlic
4 cups fresh basil or 1 cup dried basil
1 cup fresh parsley
6 tbs. raw nuts (pine, almond, hazelnut, pumpkin—I use a combination and soak them overnight)

1 cup or more of olive oil
$^1/_2$ tsp. Real Salt™ or alternative
$^1/_2$ tsp. pepper
2 tbs. sun-dried tomatoes

Combine all ingredients in a food processor (with an S-blade) or in a blender. Blend until smooth.

Maren's Salsa

SERVES 6–8

6 cloves minced garlic

Finely chop:
1 yellow onion
$^1/_2$ cup fresh coriander
$^1/_2$ cup fresh parsley
7 ripe tomatoes
1 green pepper
1 red pepper

Juice of 1 lemon and 1 lime (about 10 tbs.)
Real Salt™ or alternative to taste
Mexican seasoning to taste
Cajun seasoning (by Tones), to taste
Cayenne pepper, to taste
Cumin, to taste

Blend together and refrigerate.

Tomato Gravy

SERVES 6–8

1¹/₂ pints/800 ml. peeled tomatoes, puréed
1 small aubergine, diced
6 tbs. finely chopped green pepper
3 tbs. olive oil
Real Salt™ or alternative to taste

Cook puréed tomatoes in saucepan on medium heat. After 15 minutes of cooking the purée, stir in aubergine and pepper and shut off heat. When cooled slightly, add remaining ingredients and blend. Serve warm on raw or steamed courgettes.

Tomato Sauce

SERVES 4–6 (MAKES ABOUT 3¹/₂ CUPS)

¹/₂ cup onion, chopped
¹/₂ cup vegetable stock
3 cups tomatoes, coarsely chopped
¹/₂ tsp. oregano
¹/₂ tsp. thyme
¹/₂ tsp. basil
1 tsp. garlic powder
Freshly ground pepper

Cook onion in stock until soft. Add remaining ingredients. Bring to a boil, cover, and simmer 30–45 minutes. Add other herbs and spices to flavor as desired. Store in a glass jar and refrigerate until ready to use. For good food combination, eat with celery, bell peppers, cucumbers, aubergine, okra, or summer squash.

Italian Tomato Sauce

SERVES 14 ($\frac{1}{2}$-CUP SERVINGS)

2 28-oz./800g. tins Italian tomatoes (crushed)
1 tsp. basil
$\frac{1}{2}$ tsp. oregano
1 6-oz./150g. tin tomato paste
1 bay leaf
5 tsp. garlic (minced)
$\frac{1}{2}$ tsp. cayenne pepper

Mix all ingredients; simmer on stove for two hours. Use in your favorite Italian recipe or serve over spaghetti.

CONDIMENTS, DIPS, SPREADS, AND FILLINGS

Try thinning the spread recipes with veggie juice, water, or oil and using them as salad dressings.

Mock Mayo

Serves 4

1 cup steamed cauliflower
¼ tsp. Real Salt™ or alternative
¼ tsp. dry mustard
¼ tsp. paprika
½ tsp. powdered lecithin (optional)
½–1 tsp. linseed oil (Udo's) to emulsify (optional)
Cayenne pepper, to taste

Whip cauliflower with a little water and linseed oil, and add the remaining ingredients or process in a food processor until creamy and smooth.

Mock Almond Mayonnaise

SERVES 4

1 cup soaked almonds
$^{1}/_{2}$ cup water or veggie broth
1 lemon, peeled and chopped
1 tbs. dried onion or 3 tbs. chopped onion
3 tbs. chopped red pepper
1 clove garlic
1 tbs. oil (Udo's Choice® or linseed oil)
1 tsp. dried oregano or 1 tbs. fresh
2 tsp. dulse flakes
2 tsp. Bragg™ Liquid Aminos
Pinch each of cumin, curry seasoning, and Zip (Spice Hunter) or alternative

In a food processor or blender combine the almonds with water or broth and process until smooth. Add lemon, onion, red pepper, garlic, oil, oregano, Liquid Aminos, dulse flakes, and spices. Blend until smooth, using additional water if necessary to achieve the desired consistency. This can be a great dressing for salad or a dip with dehydrated veggies added. Enjoy!

AvoRado AvoCado Topping

SERVES 6

This is a great topping that is all raw and will work well with various seasonings of your choice. It is great as a veggie dip or wonderful when spooned on top of the Millet/Buckwheat Pancakes (see page 334).

1 cup raw almonds, ground to a fine powder
¹/₂ cup oil (I use olive)
4 tbs. water
1 avocado
2 tbs. lemon or lime juice (I use both)
¹/₂–1 tsp. Mexican seasonings or alternative
¹/₂–1 tsp. Real Salt™ or alternative
¹/₃ cup red pepper, finely chopped
¹/₃ cup red onion, finely chopped
1 carrot, finely grated

Grind almonds to a powder, then add oil, water, avocado, lemon or lime juice, salt, and seasonings of choice. Mix well in blender, coaxing batter down the sides with a spatula. Take out of blender and put in a bowl.

In a food processor finely chop the red pepper and red onion, then put the grating blade on and grate the carrot into the red pepper and onion mixture. Stir all together and then spoon into the avocado mixture in the bowl. Stir until well combined and chill before serving.

Ginger-Almond Paste Topping

SERVES 2–3

This recipe is similar to the Wowie Zowie Almond Butter Dressing (see page 267), but is much thicker. Try it on Millet Yam Hash Browns (see page 332) hot out of the oven.

¹/₂ cup almonds (I used ¹/₂ cup almond butter in a pinch, or you could use other nuts, such as macadamias or pecans)

$^{1}/_{4}$–$^{1}/_{2}$ tsp. Real Salt™ or alternative to taste
Juice of 1 lemon
1 tbs. minced fresh ginger
$^{1}/_{2}$ tsp. dried onion
1 clove garlic, minced (could be roasted for a nice flavor
    difference)

Put all ingredients into a food processor and let it fly! Add
water to thin if desired.

Basic Seasoning Recipe

SERVES 6–8

1$^{1}/_{2}$ oz./40g. onion powder
$^{1}/_{2}$ oz./10g. garlic powder
2 oz./50g. comfrey leaf or celery leaf powder or mix
$^{1}/_{2}$ tsp. red cayenne pepper
$^{1}/_{2}$ tsp. Real Salt™ or alternative
$^{1}/_{2}$ oz. ginger root powder

Mix all ingredients together. Store in a tightly capped jar
and use as a vegetable seasoning.

Herb Oil

Serves 4 (makes ¾ cup)

½ **cup linseed, olive, Essential Balance, or Udo's oil**
2 tbs. lemon juice
½ **tsp. Real Salt™ or alternative**
⅛ **tsp. freshly ground black pepper**
¼ **cup finely chopped fresh parsley**
½ **tsp. dried tarragon leaves**
Dash of cayenne pepper

Mix all ingredients well. Store in jar, refrigerate. Great on salads and steamed veggies.

Green Pepper Relish

Makes 5 cups

6 large green peppers
2 small fresh hot chili peppers
Linseed oil
Real Salt™ or alternative to taste
Cumin, to taste

In 450°F/230°C/gas 8 preheated oven place green peppers and chili peppers in shallow baking dish. Bake 20 minutes, turning once. Put peppers in pan of cold water, skin them, and remove seeds. Chop peppers very fine. Add enough linseed oil to give spreading consistency. Season with salt and cumin to taste. For sweeter relish use red peppers instead of green. Store in jar with tight lid in the refrigerator.

Chickpea Spread

SERVES 6–8 (MAKES 3 CUPS)

2 cups sprouted or tinned chickpeas
1 chopped medium onion
2 tbs. dried parsley
1 tsp. Real Salt™ or alternative
1 tsp. coriander
Dash cayenne or chili powder
¹/₄ cup water

Blend all in blender until smooth. Spread on sprouted whole wheat tortillas (better to use with veggies). Top with alfalfa sprouts or eat with veggies.

Zippy Chickpea Spread

SERVES 6–8 (MAKES 3 CUPS)

4 cups sprouted, cooked chickpeas
3 tbs. tahini
3 lemons or limes
5–6 cloves of fresh garlic, pressed
1 medium onion, chopped
2 tbs. dried parsley
Dash of cumin
1 tsp. Real Salt™ or alternative
1 tsp. coriander
Dash cayenne or chili powder
¹/₄ cup water

Blend all in blender until smooth. Spread on sprouted whole wheat tortillas, topped with alfalfa sprouts, or eat with veggies.

Guacamole

SERVES 2

1 large ripe avocado
1 tomato, finely chopped
$1/4$ tsp. Real Salt™ or alternative
$1/8$ cup lime or lemon juice
Chili powder to taste

Mash avocado and mix with other ingredients. Use as a salad dressing or serve as dip for raw bell peppers, celery, aubergine, cucumber, or summer squash.

Fresh Spinach Filling

SERVES 6

$1 1/4$ cups finely chopped fresh spinach
3 tbs. Mock Mayo (see page 349)
1 tbs. chopped pimento
$1/4$ tsp. onion powder

Combine all ingredients and mix well. Season to taste. Delicious in sprouted wheat tortillas.

Crispy Radish Filling

SERVES 6

$^3/_4$ cup finely chopped celery
$^1/_2$ cup finely chopped radishes
4 tbs. Mock Mayo (see page 349)
1 tbs. chopped chives
$^1/_4$ tsp. Real Salt™ or alternative
Few grains of pepper

Combine all ingredients and mix well. Great for stuffing celery, on sprouted wheat bread, or on sprouted wheat tortilla roll-ups.

Garden Variety Filling

SERVES 6

$^3/_4$ cup grated carrot
$^1/_2$ cup finely chopped celery
2 tbs. grated flavoured soy cheese
3 tbs. Mock Mayo (see page 349)
1 tbs. finely chopped green pepper
1 tbs. Bragg™ Liquid Aminos
$^1/_4$ tsp. Real Salt™ or alternative
$^1/_4$ tsp. cayenne pepper
Few grains of black pepper

Combine all ingredients and mix well. Excellent for stuffing celery sticks or to eat with vegetables.

Tofu/Avocado Dip

SERVES 6

1 pack soft *fresh* tofu, drained
1¹⁄₂ tsp. lemon juice
1 tsp. garlic powder
1 tbs. diced onion
2 tbs. chopped fresh coriander
¹⁄₂ tsp. chili powder
1 small tomato, diced (optional), or 2–3 sun-dried
   tomatoes
1 medium avocado, mashed
¹⁄₂–1 tsp. Real Salt™ or alternative

In a blender or food processor, combine tofu, lemon juice, garlic
powder, onion, fresh coriander, and chili powder and process
until well blended. Put the mixture in a bowl, add tomato,
avocado, and salt, and mix well. Or put all ingredients in
food processor. Process until smooth and serve chilled with
chips or fresh veggies.

Tahini (Sesame Seed Butter)

SERVES 6

1 cup sesame seeds
2 tsp. linseed oil (or preferred oil)

Combine in food processor or blender. Blend into a smooth
paste. This is a protein food, for use with nonstarchy vegetables. Use up quickly. Keep refrigerated, tightly covered.

Great Olé Guacamole

SERVES 6–8

Use this as a dip for fresh veggies. Cut the veggies like bell peppers and cabbage with small cookie cutters for children.

Juice of 1 lemon or lime (or use both)
1 large tomato
3 avocados
1 tsp. Mexican seasoning
¼ tsp. cumin
½ tsp. Zip (Spice Hunter) or alternative
1 tsp. Real Salt™ or alternative to taste

In a food processor fitted with an S-blade, blend the lemon juice with half the tomato and half the avocado until smooth. Finely dice the remaining half tomato. Mash the remaining avocado, leaving the pulp chunky. Combine both mixtures, seasoning with the Mexican Seasoning, cumin, Zip, salt, and additional lemon juice to taste, if desired. Olé!

Zippy Coriander Dip

SERVES 6–8

½ cup chopped fresh coriander
1 or 2 hot chili peppers
2 cups frozen *petits pois*, thawed
1 pack *fresh* tofu, drained
1 tbs. lemon juice
1 tsp. ground cumin

¼ tsp. freshly ground pepper
Real Salt™ or alternative to taste
1 medium cucumber

Combine one-quarter of the cilantro and all other ingredients except the cucumber in a food processor and process until smooth, approximately 30 seconds on high. Refrigerate for an hour. Lay overlapping thin cucumber slices and rim with the remaining coriander. Serve with raw vegetables.

Leprechaun Surprise Dip

SERVES 6–8

2 cups spinach, very finely chopped
2 cups parsley, very finely chopped
1 cup green onions, very finely chopped
½ cup Mock Mayo (see page 349)

Mix well. Serve with fresh vegetables.

Hearty Nut Filling

SERVES 6

½ cup almond butter
¼ cup finely chopped green pepper
¼ cup grated carrot
1 tsp. minced onion
4 tbs. Mock Mayo (see page 349)
1½ tsp. salt
½ medium red onion, finely chopped

Mix all ingredients, put in food processor if desired. Excellent for stuffing celery sticks or with vegetables.

Raw Pecan Paté

SERVES 8

This pâté spreads well on tortillas and celery. The fresh raw pecans and shredded veggies make it a sweet pâté that's great for children!

2 cups fresh raw pecans
$^1/_4$ –$^1/_2$ red onion
4–6 fresh basil leaves
$^1/_4$ cup finely grated carrots, beetroots, and/or raw
     squash
$^1/_4$ cup finely minced parsley (optional)

In a food processor fitted with an S-blade, blend the pecans, onions, and basil leaves. Thin with enough water (optional) to desired consistency like pâté. Add the grated veggies and keep blending until well mixed and moist. Stir in the minced parsley and mix well. You can even make this into patties and warm in a food dehydrator to the desired warmth and crispness (four to eight hours) if you wish. If you are in a hurry, you could warm the pâté lightly in a skillet right before you serve it. Other spices can be added to the pâté too. Experiment!

Almond Pâté

SERVES 8–10

3 cups soaked almonds
1 cup lemon juice
¼ cup Bragg™ Liquid Aminos
½–1 clove garlic (could use roasted garlic too)

In food processor, process the almonds, lemon juice, Liquid Aminos, and garlic until smooth. Store in an airtight container in the fridge.

Variations: This is a basic pâté spread recipe. You can vary it by using pine nuts, sesame seeds, or other nuts, such as soaked hazelnuts or pecans. You can cream it up more by adding tahini. Also try to season it differently with various spices. Fresh herbs or dehydrated veggies can color and flavor it up. Be daring and creative! Use to stuff peppers or celery. Spread on crackers or wraps.

Tofu Pâté

SERVES 6–8

1 lb./450g. firm *fresh* tofu, drained
1 tbs. Bragg™ Liquid Aminos
1 tbs. sesame tahini
1 tbs. linseed oil
2 tbs. yeast-free vegetable broth
1 tbs. minced chives
1 tbs. minced fresh basil

Place all ingredients in a mixing bowl and mix thoroughly until smooth. Press the mixture into a mold and refrigerate for two hours.

Slice or scoop out and serve with nonstarchy vegetables.

Green Mayonnaise

Serves 6–8

1 lb./450g. tofu, drained
2 avocados
½ tbs. curry powder
3 tbs. lemon juice
Salt, to taste
Zip (Spice Hunter)or alternative to taste

Place all ingredients in a food processor and process until smooth and creamy.

Guacamole Green Mayonnaise Variation

Serves 4

1 tomato, chopped, peeled
½ onion, chopped, peeled
2–3 seeded green chili peppers

Blend all ingredients in a processor until mayonnaise consistency.

To make this Guacamole variation, add these ingredients to the ingredients for the Green Mayonnaise recipe (see above).

Variation: Add parsley, chives, tarragon, or other spices of choice.

Yummus Hummus

Serves 6–8

2–3 tbs. olive oil
Juice of 1 lemon
1–2 garlic cloves, minced
$^1/_8$–$^1/_4$ cup tahini (raw)
1 17-oz./450g. jar chickpeas, drained (save the water)
Real Salt™ or alternative to taste
$^1/_2$–1 tsp. garlic herb seasoning or alternative
$^1/_2$–1 tsp. cumin
Zip (Spice Hunter) or alternative to taste

In a food processor put oil, lemon juice, garlic, and tahini and process until smooth.

Add chickpeas and seasonings and process until creamy.

You may need to thin with extra water (from chickpeas) to desired consistency. Serve on wraps, with raw veggies, or in pitta sandwiches.

Great variations: Add an avocado to this recipe to make the hummus green and creamier, or add one or two red or orange bell peppers to make the hummus sweeter and more vibrant orange. Add a few sun-dried tomatoes for a more flavorful and deeper-colored hummus. Experiment and enjoy!

Sweet Carrot Butter

SERVES 8–10

This is a sweet, creamy spread that is wonderful on wraps or used as a dip for raw veggies. Especially nice when something sweet is called for. Kids like this one too.

2 cups raw macadamias
2–3 grated organic carrots
Vanilla to taste
Olive oil and water to thin

In a food processor, mix nuts and carrots and process until smooth and creamy. Add a few drops of vanilla while processing if desired. While processor is running, add olive oil and/or water to desired consistency.

Sauce Sampler Platter

SERVES 8–12

This is a great way to serve a variety of tastes and textures. Makes a great appetizer or can be used as a main meal for any time. It's wonderful for parties too!

Roasted Pepper Macadamia Sauce (see page 308)
Sweet Carrot Butter (above)
Spring's Pesto (see page 345)
Yummus Hummus (see page 363)
Maren's Salsa (see page 346)
Raw veggies of your choice (for example, carrot sticks,

bell pepper strips, broccoli and cauliflower florets, raw
yam slices, cucumbers, arrowroot sticks)
Baked tortilla chips
Crackers

On a plate scoop a big mound of each of the sauces, and
serve with veggies, crackers, and chips for dipping.

JUICES

Basic Green Vegetable Juice

Serves 1

This is a power-packed green drink. Go easy on the parsley as it has a very strong flavor. Instead of this you can drink 1 tsp. concentrated green powder in 8 oz./225 ml. pure water.

2–3 stalks celery
1 cucumber
2–3 large leaves kale
4–5 large leaves lettuce
2 cups spinach
$^1/_4$–$^1/_2$ cup parsley

Vegetable/Grass Drink

Serves 1

1–3 oz./25–75 ml. carrot juice (3 carrots or less, try to keep the drink at least 80 percent green)
3 oz./75 ml. celery juice (2 large stalks)
$^1/_2$ oz./14 ml. parsley juice (5 sprigs)
1$^1/_2$ oz./50 ml. wheat grass juice

Wheat-Beet Juice

SERVES 1

1¹/₂ oz./50 ml. wheat grass juice
1 oz./25 ml. beetroot juice
6 oz./175 ml. cucumber juice

Basic Green Drink

SERVES 2

4 cups alfalfa and/or other sprouts
4 cups sunflower and buckwheat greens
¹/₂ cup carrots
¹/₂ cup sweet red peppers
¹/₄ cup parsley
1 cup cucumber
Add 1 bunch wheat grass (about ³/₄ inch thick), if desired

Garden Green Drink

SERVES 2

4 cups sprouts
4 cups green tops
2 cups kale or collard greens
1 cup celery

Green Power Cocktail

SERVES 2

4 cups sprouts
4 cups green tops
1 cup kale
1 cup beetroots
$^1/_2$ cup wheat grass

Spring Green Drink

SERVES 2

4 cups sprouts
4 cups greens
$^1/_2$ cup dandelion greens
$^1/_4$ cup spring onions
1 cup carrots

Potassium Special

SERVES 2

3 oz./75g. carrots
4 oz./100g. celery
2 oz./50g. parsley
3 oz./75g. spinach

Insulin Generator

SERVES 2

3 oz./75g. Brussels sprouts
3–6 oz./75–175g. carrots
3 oz./75g. string beans
4 oz./100g. lettuce

High Vitamin C & E Drink

SERVES 2

6 oz./175g. spinach
2 oz./50g. lettuce
2 oz./50g. watercress
4 oz./100g. carrots
2 oz./50g. green peppers

Blood Builder

SERVES 2

8 oz./225g. celery
3 oz./75g. cucumber
2 oz./50g. parsley
3 oz./75g. spinach

Skin Cleanse

SERVES 2

4 oz./100g. potato
4 oz./100g. celery
3–6 oz./75–175g. carrots
2 oz./50g. watercress

All-Vegetable Cocktail

SERVES 2

$^3/_4$ pt./450 ml. fresh tomatoes
$^1/_2$ tsp. garlic
1 cucumber, sliced
1 green pepper
Sprigs of fresh parsley
$^1/_4$ onion, sliced
2–3 lettuce leaves
$^1/_2$ tsp. ginger

Blend all ingredients in a blender on low speed.

NUT AND SEED MILKS

Almond Milk

SERVES 2

¹/₂ **cup almonds**
¹/₂ **cup pine nuts**
1 cup spring or filtered water

Soak almonds and pine nuts for 12 hours. Put in blender and pulverize. Add the cup of water gradually, while continuing to blend on high. Strain through a fine strainer or cheesecloth (you can use the almond pulp as a body scrub). This milk will keep for three to four days. It is great on hot grains such as quinoa, buckwheat groats, millet, or amaranth. We like to add some soaked almonds to our grains for "crunch." You can thin further with more water if desired.

Quick Tahini Milk

SERVES 1–2

2–4 tbs. tahini (raw)
1 cup water

Tahini is a butter made from hulled sesame seeds, usually used as a spread (see recipe on page 357). It also makes a very nutritious milk, which is high in calcium and protein. In a blender, combine 2 tbs. of tahini with water. Blend thoroughly and taste. Add additional tahini and blend again for a richer milk. This milk will keep for three to four days.

BREADS

Camper's Bread

2 cups sprouted wheat flour
4 tbs. nonaluminum baking powder
1 tbs. Real Salt™ or alternative
2 Tbs oil (olive or Udo's Choice®)
1 cup pure water

Mix dry ingredients, cut in oil, add water, and mix well. Grease frying pan, pour in batter, cook very slowly. Turn.

Essene Bread

Serves 2

1½ pints/800 ml. sprouted grain
⅔ cup pure water

Add water to grain and grind up in a blender. Form into a small loaf and bake at 275°F/140°C/gas 1 for three hours or until crust forms. Very moist.

Sprouted Wheat Bread

SERVES 1–2

2 cups wheat

Sprout 2 cups wheat for two days, then grind.

Make a pad ⅛ inch thick from this dough.

Bake on a flat stone in full summer sun from morning until noon on one side, from noon until evening on the other side, *or* if weather is bad, bake it in a slow oven (275°F/140°C/gas 1) until slightly crisp.

This recipe is adapted from the Dead Sea Scrolls and is the same type of bread Jesus broke at the Last Supper. This is a tasty, hard tack-like bread.

CRUNCHY SNACKS

Veggie Crunch Stix and Crackers

SERVES 8–10

These colorful snacks are a great way to wean children and adults from yeast breads. They are a wonderful grab-snack and also complement and give crunch to a vegan-based meal such as soup and salad. They are quick to make and travel well. I use the small cookie cutters and make little dinosaurs, aeroplanes, and heart crackers. Also, you can season them any way you want by adding a couple of teaspoons of your favorite spice.

2 cups flour (all-purpose, millet, whole wheat—I use half and half all-purpose and whole wheat)
$\frac{1}{2}$–1 tsp. salt
$1\frac{1}{2}$ tsp. baking powder
3 heaped tbs. soft tofu
2 tbs. olive oil
1–2 tsp. seasonings of your choice (optional)
$\frac{1}{2}$–$\frac{3}{4}$ cup cold water or fresh vegetable juice, or mix

In a food processor, pulse the flour, the salt, and baking powder to combine. Add the tofu and olive oil and pulse until the mixture resembles coarse meal. With the machine running, gradually add between $\frac{1}{2}$ and $\frac{3}{4}$ cup cold water or fresh veggie juice until the dough comes together in a soft ball (approximately one minute).

Turn the dough out onto a lightly floured surface. Form the dough into a smooth rectangle, about 4 × 6 inches/10 × 15 cm, then roll out the dough into an 8 × 10-inch/20 × 25 cm sheet, ¼ inch/5 mm thick. With a sharp knife cut the dough lengthwise into ¼ inch/5 mm wide strips.

Using your hands, gently roll each strip into 16 inch/40 cm long sticks. For a twisted version, grab each end of the dough strip with your fingers and carefully stretch and twist the strip in opposite directions. For crackers use cookie cutters (children love these!) and arrange on baking sheet. Arrange the sticks on two baking sheets, side by side but not touching, and press ends into the baking sheet to keep the sticks straight while they cook. If desired, brush each stick lightly with olive oil and sprinkle with salt or seasonings of your choice. Bake at 350°F/180°C/gas 4 until firm and cooked through, 14–18 minutes.

Transfer the sticks or crackers to a wire rack to cool. Store in an airtight container at room temperature for two to three days.

Variations:

Beet Stix: 2 tbs. beetroot juice, from 1 small beetroot. Combine with ½ cup cold water.

Popeye Stix: ½ cup parsley or spinach juice. Combine with ¼ cup cold water.

Bugs Bunny Stix: ¼ cup carrot juice, from about three carrots. Combine with ¼ cup cold water.

Tomato Stix: ¼ cup fresh tomato juice, with 1–2 tbs. sun-dried tomato pesto. Combine with ⅓ cup cold water.

Spicing Variations:

Curry/Turmeric Stix: 1 tsp. curry powder, with ½ tsp. ground turmeric.

Cumin Stix: 2 tsp. ground cumin.
Garlic Stix: 2 tsp. garlic herb seasoning.
Mexi Stix: 2 tsp. Mexican seasoning.
Experiment!

Spicy Pecan Croutons

SERVES 4–6

These spicy, toasty pecan croutons will add pizzazz to any salad. That's if they make it into the salad—they usually go down as a pop-in-the-mouth snack around my house. The cayenne pepper adds magnesium and bioflavonoids to enhance circulation.

2 tbs. grape seed or olive oil
1 cup raw pecans (almonds or pine nuts also work great)
$^1/_2$–1 tsp. cayenne pepper (start with $^1/_2$ tsp. and work up if you like them hotter)
1 tsp. curry seasoning or plain ground cumin (I use cumin)
$^3/_4$ tsp. Real Salt™ or alternative

In a frying pan heat the oil over medium-low heat. Add the pecans and all other ingredients and sauté until the nuts are well coated and lightly toasted. Serve warm from the pan over salads. And serve immediately. These croutons are also wonderful sprinkled over stir-fried veggies such as asparagus or green beans. These also can be cooled and used for snacks. Enjoy!

Another great serving idea: Make the Blackened Herbed

Fillets (see page 289) and sauté some asparagus to place on top of the fish when done. Then as a garnish, sprinkle these pecan croutons over the top. Serve with a big helping of Rainbow Salad (see page 254). *Bon Appétit!*

Crispy Buckwheat Groats

SERVES 6–8

2 cups hulled buckwheat groats, soaked 6–8 hours
Bragg™ Liquid Aminos, to taste
Lemon or lime juice (I use both)
Zip (Spice Hunter) or any seasoning, to taste

Drain the buckwheat groats and put in a shallow bowl. Add Liquid Aminos and juice to cover. Add spices to taste. Soak in this solution for an hour, then drain again. Put groats in food dehydrator on teflex liners and dehydrate until dry (two to three hours).

These are great little munchy snacks and can be used as croutons in a salad or wrap.

Dehydrated Linseed Chips

SERVES 6–8

1 cup linseeds
1 tomato
½ red bell pepper
1 clove garlic

1–2 tsp. Mexican seasoning, Italian, garlic herb, or any
    other spice combo you like
$^1/_2$ small red onion or 1 tsp. dried onion flakes
$^1/_2$ beetroot
Real Salt™ or alternative to taste

Soak the linseeds in 2 cups water for two to four hours.
Process the tomato, bell pepper, garlic, Mexican spices, onion,
beetroot, and salt. Keep it chunky. Add the soaked linseeds
into the blended mixture, and spoon 2 inch/5 cm rounds
onto a teflex sheet and dehydrate at 105–110°F/40–44°C for
eight to twelve hours, or to the desired crispness. Turn the
crisps over after four to six hours to ensure even drying.

Optional: Sprinkle sesame or soaked pumpkin seeds on
top before dehydrating.

Tortilla Chips

Serves 4–8

4 large sprouted wheat tortillas, or tortillas of your
    choice
4 tbs. olive or grape seed oil, or other preferred oil
Seasonings of choice, such as sesame seeds, garlic,
    Italian, Mexican, Barbecue, or Herbes de Provence

Lay out tortillas on two nonstick baking sheets. With a
pastry brush or napkin, wipe over each tortilla so that the surface is covered with oil. Sprinkle seasoning of choice over the
top of the tortillas and bake at 350°F/180°C/gas 4 until
golden or just almost crisp, around 10 minutes.

Cool and then break into pieces for eating with dips, soups, or salads. You can also precut the tortillas with a pizza cutter before you bake them if you want a cleaner-edged cracker.

CEREAL AND DESSERT

Mock Pumpkin Pie

SERVES 6–8

This recipe was the result of trying to come up with something healthy in place of pumpkin pie at Thanksgiving. It is a healthy alternative made out of carrots. Experiment using pumpkin and other squashes.

Pie Crust:
2 cups raw almonds
2–3 tbs. soy milk or almond milk
2 tbs. wheat bran flakes

Pie Filling:
1 lb./450g. carrots, peeled and grated
$1/2$ tsp. nutmeg
$1/2$ tsp. cinnamon
$1/8$ tsp. clove
1 tsp. vanilla
$1/2$ cup slivered or whole almonds, for garnish

Pie crust: Place raw almonds in food processor and pulse-chop until fine. Add soy milk and bran 1 tbs. at a time until mixture holds together. Press into pie dish evenly.

Pie filling: Steam carrots until soft. Place in blender or food processor and process until smooth. Add spices and vanilla to taste. Pour mixture into almond crust. Garnish top of pie with slivered or whole almonds. Chill in fridge overnight and serve cold.

Sprouted Cereal

SERVES 2

2 cups wheat or rye grains (organic, unstored)
$^1\!/_2$–1 tsp. cinnamon

Soak grain overnight in distilled water. Drain and set jar on its side to sprout. Rinse sprouts morning and evening, and sprout for two days. Add enough water to sprouts to blend in blender. Pour into saucepan with cinnamon and cook until toasty warm. May be served in a bowl with soy milk.

RECIPE SUBSTITUTIONS

You can most certainly rely on many of your old favorite recipes while you are on this program. And with a few creative substitutions, even more of them will fit right in with your new way of eating. Here are the most common substitutions necessary to get you started:

If a Recipe Calls For	Substitute
1 pack dried yeast	1 tsp. nonaluminum baking powder
Whole wheat or white flour	Try spelt, buckwheat, millet flour, or a combination thereof
Milk	Soy milk, rice milk, almond milk, sesame milk (unsweetened)
Vinegar	Lemon or lime juice
Soy sauce or tamari	Bragg™ Liquid Aminos
Regular cooking oils or salad oil	Good oils such as olive, sunflower, linseed, borage, almond, grape seed, or Udo's Choice®
Butter or margarine	Good oils, as above
Cheese	Sprouts (obviously, this doesn't work in every instance, for example, when the cheese is to be melted)
Meat	Soy products (make sure they contain no yeast)

If a Recipe Calls For	Substitute
Eggs	Egg replacer (as directed on package)
Salt	Real Salt™ or Celtic Sea Salt™ or salt with dehydrated veggies
Walnuts or cashew nuts	Almonds, hazelnuts, pecans, pine nuts
White rice	Organic brown rice, brown or natural white basmati rice, spelt, buckwheat groats, millet, kamut, quinoa, amaranth
Bread	Unleavened, yeast-free, sprouted breads
Pasta	Vegetable, spelt, or artichoke pasta

Resources

Medical

For referrals for live blood analysis and the Mycotoxic/ Oxidative Stress Test (M/OST), contact the Inner Light Biological Research Center. This is also the place to call if you need information about products used in this book that this Resources section does not provide.

InnerLight Biological Research Center
134 E. 200 No.
Alpine, UT 84004
801-756-7850
www.innerlightfoundation.org

The Cutting Edge Catalogue carries many items in the category of health technology, including pH strips, water systems, and books.

Cutting Edge Catalogue
P.O. Box 5034
Southampton, NY 11969
Orders or catalogue: 800-497-9516
Information: 516-287-3813
Fax: 516-287-3112
www.cutcat.com
E-mail: cutcat@i-2000.com

Kitchen Equipment and Food

Some of the foods, supplements and equipment mentioned in this book may not be easy to find in high street health shops. However, there are many suppliers of specialist products that can be found on the Internet. If you do not have Web access at home, many local libraries now have Internet facilities and can help you find what you want.

The websites below can supply some of the products recommended in this book and are interesting to browse.

www.acorn-supplements.co.uk
(Supplies Udo's oils.)

www.amazon.co.uk
(Supplies most of the books listed in Recommended Reading.)

www.clearsprings.co.uk

www.cutco.co.uk
The Cutting Edge
12 Lion and Lamb Walk
Farnham

Surrey GU9 7LL
Tel: 0800 834320
(UK distributor of Cutco knives.)

www.foodfirst.co.uk
(Directory of producers, wholesalers, retailers and service
providers.)

www.fresh-network.com
The Fresh Network
PO Box 71
Ely
Cambridgeshire CB7 4GU
Tel: 0870 800 7070
Fax: 0870 800 7071
E-mail: info@fresh-network.com
(Supplies many products including: Green Power Juicers,
Vita-Mix blenders, Saladacco machines, Excalibur
Dehydrators, sprouting kits and seeds, Udo's oils, pH
book/testing kit, Japanese foods, almond butter, various
grains, pulses and other organic food products.)

www.goodnessdirect.co.uk
(Supplies dairy-free pesto.)

www.greenfoods.co.uk

www.health-store.co.uk
(Supplies Udo's oils, Source Natural products, Essiac tea.)

www.health4youonline.com
(Supplies Essential Balance oils.

www.homeopath.co.uk
(Supplies black haw bark, Aerobic Oxygen, hydrogen peroxide.)

www.lakelandlimited.co.uk
(Supplies kitchen equipment including cups and measuring spoons)

www.mountfuji.co.uk
Mount Fuji International Ltd
Felton Butler
Shrewsbury SY4 1AS
Tel: 01743 741169
Fax: 01743 741650
Email: info@mountfuji.co.uk
(Supplies rice cookers and Japanese foods and equipment)

www.positivehealthshop.co.uk
(Supplies Udo's and Essential Balance oils.)

www.sunnyfields.co.uk
(Supplies Bragg™ Liquid Aminos, kuzu root, soy cheeses, egg replacer, sprouted grain breads and other health/organic foods.)

Supplements

For pH drops, green powder, and more:
Inner Light Foundation
801–756–7850
www.innerlightfoundation.com

Recommended Reading

Earl Mindell's Soy Miracle
Earl Mindell
Simon & Schuster
New York, New York

Enzyme Nutrition
Edward Howell
Avery Publishing Group, Inc.
Wayne, New Jersey

Fats That Heal—Fats That Kill
Udo Erasmus
Alive Books
Burnaby, British Columbia, Canada

The HarperCollins Illustrated Medical Dictionary
HarperCollins
New York, New York

The Healing Power of Chlorophyll from Plant Life
Bernard Jensen
Bernard Jensen Enterprises
Escondido, California

A Holistic Protocol for the Immune System
Scott J. Gregory
Tree of Life Publications
Joshua Tree, California

Is This Your Child's World?
Doris J. Rapp, M.D.
Bantam Books
New York, New York

The Juicing Book
Stephen Blauer
Avery Publishing Group, Inc.
Garden City Park, New York

Prescription for Nutritional Healing
J. F. Balch and P. A. Balch
Avery Publishing Group
Garden City Park, New York

Reclaiming Our Health
John Robbins
HJ Kramer, Inc.
P.O. Box 1082
Tiburon, California 94920

Slow Burn
Stu Mittleman
HarperCollins
New York, New York

Touch for Health
John F. Thie, D.C.
DeVorss & Co.
Marina del Rey, California

Wheatgrass Book
Ann Wigmore
Avery Publishing Group, Inc.
Wayne, New Jersey

The Yeast Connection, A Medical Breakthrough
William G. Crook
Professional Books
Jackson, Tennessee

References

Adetumbi, M. A., Javor, C. F., and Lau, B. H. S. Anti-Candida activity of garlic—effect on macromolecular synthesis. Presented at the American Society for Microbiology, Loma Linda University, 1985.

Alberts, B., et al., eds. *Molecular Biology of the Cell,* 2d ed. New York: Garland Publishing, Inc., 1989.

Aleksandrowixz, J., and Smyk, B. Mycotoxins and their role in oncogenesis with special reference to blood diseases. *Polish Medical Science Historical Bulletin,* 1971; 24: 25–30.

Alexander, J. G. Allergy in the gastrointestinal tract. *Lancet,* 1975; 2: 1264.

Alpert, M. E., Hutt, M. S. R., Wogan, G. N., and Davidson, C. S. Association between aflatoxin content and hepatoma frequency in Uganda. *Cancer,* 1971; 28: 253.

Aso, H., et al. Induction of interferon and activation of NK cells and macrophages in mice by oral administration of Ge-132, an organic germanium compound. *Journal of Microbiology and Immunology,* 1985; 29(1): 65–74.

Béchamp, Pierre Jacques Antoine. *The Blood and Its Third Anatomical Element* (Montague R. Leverson, translator). London: John Ouseley Limited, 1912.

Becker, Robert O., M.D., and Selden, Gary. *The Body Electric. Electromagnetism and the Foundation of Life.* New York: Quill/William Morrow, 1985.

Bertz, A., et al. Modulation by cytokines of leukocyte endothelial cell interactions. Implications for thrombosis. *Biorheology,* 1990; 27: 455.

Bick, R. L. Disseminated intravascular coagulation. *Hematology/Oncology Clinics of North America,* 1993; 6: 1259.

Bird, Christopher. *Gaston Naessens.* Tiburon, Calif.: H. J. Kramer, Inc., 1991.

———. *The Galileo of the Microscope.* St. Lambert, Quebec, Canada: Les Presses de l'Université de la Personne, Inc., 1990.

———. To Be or Not to Be? A paper presented in an address to L'Orthobiologie Somatidienne Symposium 1991, Sherbrooke, Quebec, hosted by Gaston Naessens.

Blank, F. O., Chin, G., Just, B., et al. Carcinogens from fungi pathogenic for man. *Cancer Research,* 1968; 28: 2276.

Bleker, Dr. Maria. *Blood Examination in Darkfield According to Professor Dr. Günther Enderlein.* Gesamtherstellung, Germany: Semmelweis-Verlag, 1993.

Boeing, H., Schlehofer, B., Blettner, M., Wahrendorf, J. Dietary carcinogens and the risk for glioma and meningioma in Germany. *International Journal of Cancer,* 1993; 53(4): 561–65.

Bolton, S., and Null, G. The medical uses of garlic: Fact and fiction. *American Pharmacy,* August 1982.

Bowie, E. J., et al. The clinical pathology of intravascular coagulation. *Bibliotheca Haematologica,* 1983; 49: 217.

Bredbacka, S., et al. Laboratory methods for detecting disseminated intravascular coagulation (DIC): New aspects. *Acta Anaesthesiologica Scandinavica,* 1993; 37: 125.

Breen, F. A., et al. Ethanol gelation: A rapid screening test for intravascular coagulation. *Annals of Internal Medicine,* 1970; 69: 1197.

Burkitt, D. Some disease characteristics of modern Western civilization. *British Medical Journal,* 1973; 1: 274.

Carp, H., et al. In vitro suppression of serum elastase-inhibitory capacity by ROTS generated by phagocytosing polymorphonuclear leukocytes. *Journal of Clinical Investigation,* 1979; 63: 793.

Chandler, W. L., et al. Evaluation of a new dynamic viscometer for mea-suring the viscosity of whole blood and plasma. *Clinical Chemistry,* 1986; 32: 505.

Chen, F., Cole, P., Mi, Z., Xing, L. Y. Corn and wheat-flour consumption and mortality from esophageal cancer in Shanxi, China. *International Journal of Cancer,* 1993; 4(2): 163–69.

Cho, T. H., et al. Effects of *Escherichia coli* toxin on structure and permeability of myocardial capillaries. *Acta Pathologica Japonica,* 1991; 41: 12.

Colucci, M., et al. Cultured human endothelial cells: An in vitro model of vascular injury. *Journal of Clinical Investigation,* 1983; 71: 1893.

Cooper, L. A., and Gadd, G. M. Differentiation and melanin production in hyaline and pigmented strains of *Microdochium bolleyi.* In Constantini, A. V., Weiland, H., Qvick, Lars I. *The Fungal/Mycotoxin Etiology of Human Disease,* Vol. 2. Freiburg, Germany: Johann Friedrich Oberlin Verlag, 1994.

Cope, Freeman W. Evidence from activation energies for superconductive tunneling in biological systems at physiological temperatures. *Physiological Chemistry and Physics,* 1971; 3: 403–10.

Costantini, A. V., Weiland, H., Qvick, Lars I. *The Fungal/Mycotoxin Etiology of Human Disease.* Volumes 1 and 2. Freiburg, Germany: Johann Friedrich Oberlin Verlag, 1994.

Cusumano, V. Aflatoxin in patients with lung cancer. *Oncology,* 1991; 48: 194–95.

Dickens, L. *Carcinogenesis: A Broad Critique.* Baltimore: Williams & Wilkins, 1967.

Dickens, R., and Jones, H. E. H. Further studies on the carcinogenic action of patulin-induced mammary adenomas and local sarcomas or fibrosarcomas in mice and rats. *British Journal of Cancer,* 1965; 19: 392.

Duke, Don, M. S. Materials rich in monoatomic elements [report on personal research]. Phoenix, Ariz., 1995.

Encyclopedia of Chemical Technology. New York: John Wiley and Sons, 1983.

Enderlein, Prof. Dr. Günther. *Akmon,* Volume I, Books 1 and 2. Hamburg, Germany: Ibica-Verlag, 1957.

———. *Bakterien Cyclogenie.* Hamburg, Germany: Ibica-Verlag, 1925.

Enomoto, M. Carcinogenicity of mycotoxins. In *Toxicology, Biochemistry and Pathology of Mycotoxins* (Uraguchi, K., and Yamazaki, M., eds.). New York: John Wiley & Son, 1978.

Fink-Gemmels, J. The significance of mycotoxin assimilation of meat animals. *Deutsche Tierärztliche Wochenschift,* 1989; 96(7): 360–63.

Franceschi, E. A. Meat, poultry, cooked ham, salami, sausages, cheese, butter and oil-related thyroid cancer. *International Journal of Cancer,* 1993; 53(4): 561–65.

Fungalbionics Convention: The Fungal/Mycotoxin Etiology of Chronic and Degenerative Disease. Metro Toronto Convention Centre, September 30, 1994.

Ghadirian, P. Thermal irritation and esophageal cancer in northern Iran. *Cancer,* 1987; 60(8): 1909–14.

Giovannucci, E., Rimm, E. B., Colditz, G. A., Stampfer, M. J., Ascherio, A., Chute, C. C., and Willett, W. C. A prospective study of mycotoxins and risk of prostate cancer. *Journal of the National Cancer Institute,* 1993; 85(19): 1538–40.

Grimstad, I. A., et al. Thromboplastin release, but not content, correlates with spontaneous metastasis of cancer cells. *International Journal of Cancer,* 1988; 41: 427.

Gunji, Y., et al. Role of fibrin coagulation in protection of murine tumor cells from destruction by cytotoxic cells. *Cancer Research,* 1988; 48: 5216.

Hamilton, P. J., et al. Disseminated intravascular coagulation: A review. *Journal of Clinical Pathology,* 1978; 31: 609.

Hay, E. D., ed. *Cell Biology of Extracellular Matrix.* New York: Plenum Press, 1981.

Heinicke, R. M. The pharmacologically active ingredient of noni [a paper]. University of Hawaii, January 1996.

Hertog, M. G., Feskens, E. J., Hollmati, P. C., Katan, M. B., and Kromhout, D. Dietary antioxidants and risk of coronary disease. *Lancet,* 1993; 342: 32–34.

Hills, Christopher. *Nuclear Evolution.* Boulder Creek, Calif.: University of the Trees Press, 1977.

Hu, T., et al. Synthesis of tissue factor messenger RNA and procoagulant activity in breast cancer cells in response to serum stimulation. *Thrombosis Research,* 1993; 72: 155.

Hudson, David. Alchemical research: DNA alteration and the rediscovery of the light of life. Yelm, Wash.: Leading Edge Research; Article 79, February 1995.

Hume, E. Douglas. *Béchamp or Pasteur? A Lost Chapter in the History of Biology,* 1st ed. Ashingdon, Rochford, Essex, England: The C. W. Daniel Company, 1923; 2d. ed. (London: C. W. Daniel Company, 1932) reprinted by Health Research: Pomeroy, Wash., 1989.

Hunder, G., Schumann, K., Strugala, G., Gropp, J., Fichtl, B., and Forth, W. Influence of subchronic exposure to low dietary deoxynivalenol, a trichothecene mycotoxin, on intestinal absorption of nutrients in mice. *Food Chemistry Toxicology,* 1991; 29(12): 809–14.

Ingram, D. M., Nottage, E., and Roberts, T. The role of *Saccharomyces cerevisiae*—baker's, or brewer's, yeast—in the development of breast cancer: A case-control study of patients with breast cancer, benign epithelial hyperplasia and fibrocystic disease of the breast. *British Journal of Cancer,* 1991; 64(1): 187–91.

Iwata, K., ed. *Yeasts and Yeast-Like Micro-Organisms in Medical Science.* Tokyo: University of Tokyo Press, 1976.

Jones, T. W. Observations on some points in the anatomy, physiology, and pathology of the blood. *British Foreign Medical Review*, 1842; 14: 585.

Jonsyn, Lahai. Aspergillus/aflatoxin contamination of dried fish. *International Journal of Cancer*, 1991; 4(1): 8–11.

Kalokerinos, A., and Dettman, G. *Second Thoughts About Disease: A Controversy and Béchamp Revisited*. Warburton, Victoria, Australia: Biological Research Institute [booklet published from an article in *Journal of the International Academy of Preventive Medicine*, July 1977; 4(1): 18].

Keys, A. The role of the diet in human atherosclerosis and its complications. In *Atherosclerosis and Its Origin* (Sandler, M., and Bourne, G. H., eds.). New York and London: Academic Press, 1963.

Kikuchi, S., Okamoto, N., Suzuki, T., Kawahara, S., Nagai, H., Sakiyama, T., Wada, O., and Inaba, Y. A case-control study of breast cancer/mammary cyst and dietary, drinking or smoking habits in Japan. *Japanese Journal of Cancer Clinics*, 1990; 24: 365–69.

Kono, S., Imanishi, K., Shinchi, K., Yanai, F. Relationship of diet to small and large adenomas of the sigmoid colon. *Japan Journal of Cancer Research*, 1993; 84(1): 9–13.

Kwon-Chung, K. J., and Bennet, John E. *Medical Mycology*. Malvern, Penn.: Lea and Febiger, 1992.

La Vecchia, C., Decarli, A., Negri, E., Parazzini, F., Gentile, A., Cecchetti, G., Fasoli, M., and Franceschi, S. Dietary factors and the risk of epithelial ovarian cancer. *Journal of the National Cancer Institute*, 1987; 79(4): 663–69.

La Vecchia, C., Negri, E., Decarli, A., D'Avanzo, B., and Franceschi, S. A case-control study of diet and gastric cancer in northern Italy. *International Journal of Cancer*, 1987; 40(4): 484–89.

Lancaster, M. C., Jenkins, F. P., and Philp, J. M. C. L. Toxicity associated with certain samples of broken or ground nuts. *Nature*, 1961; 192: 1095–96.

Levi, F., Franceschi, S., Negri, E., and La Vecchia, C. Dietary factors and the risk of endometrial cancer. *Cancer,* 1993; 71(11): 3575–81.

Linderfelser, L. A., Lillehoj, E. B., and Burnmeister, H. R. Aflatoxin and trichothecene toxins: Skin tumor induction and synergistic acute toxicity in white mice. *Journal of the National Cancer Institute,* 1974; 52: 113.

Livingston-Wheeler, Virginia, M.D. *The Conquest of Cancer.* New York: Franklin Watts, 1984.

Longenecker, Gesina L., Ph.D. *How Drugs Work.* Emeryville, Calif.: Ziff-Davis Press, 1994.

Lorber, A., et al. Clinical application of heavy metal complexing of N-actyl cysteine. *Journal of Clinical Pharmacology,* 1973; 13: 332–36.

Lynes, Barry. *The Cancer Cure That Worked! Fifty Years of Suppression.* Queensville, Ontario, Canada: Marcus Books, 1987.

Mackman, et al. Lipopolysaccharides—mediated transcriptional activation of the human tissue factor gene in THP-1 monocytic cells requires both activator protein 1 and nuclear factor kappa B binding sites. *Journal of Experimental Medicine,* 1991; 174: 1517.

Maier-Kopf, P. Complexes of metals other than platinum as anti-tumor agents. *Journal of Clinical Pharmacology,* 1994; 47: 1–16.

Margolis, J. The interrelationship of coagulation of plasma and release of peptides. *Annals of the New York Academy of Sciences,* 1963; 104: 133.

Margulis, Lynn, and Sagan, Dorion. *Micro-Cosmos.* New York: Summit Books, 1986.

Mattman, Lida H. *Cell Wall Deficient Forms—Stealth Pathogens.* Cleveland: CRC Press, 1974.

Miles, M. R., Olsen, L., and Rogers, A. Recurrent vaginal candidiasis; importance of an intestinal reservoir. *Journal of the American Medical Association,* 1977; 238: 1836–37.

Morrison, D. C., et al. The effects of bacterial endotoxins on host mediation systems. *American Journal of Pathology,* 1978; 93: 526.

Motola, Lynne. Hidden in plain sight, the meaning of "grass" in Hebrew. *Western Wheatgrass Journal,* January–March 1995; 2(1): 3–4.

Mueller, H. E., et al. Increase of microbial neuraminidase activity by the hydrogen peroxide concentration. *Experientia,* 1972; 23: 397.

Muller-Berghaus, G., et al. The role of granulocytes in the activation of intravascular coagulation and the precipitation of soluble fibrin by endotoxin. *Blood,* 1975; 45: 631.

Nachman, R. L., et al. Detection of intravascular coagulation by a serial-dilution protamine sulfate test. *Annals of Internal Medicine,* 1971; 75: 895.

Nachman, R. L., et al. Hypercoagulable states. *Annals of Internal Medicine,* 1993; 119: 819.

Neuhauser, I., and Gustus, E. L. Successful treatment of intestinal moniliasis with fatty acid resin complex. *Archives of Internal Medicine,* 1954; 93: 53–60.

New Frontier Newsletter. Salt Lake City: New Frontiers, Inc., November 1994.

Norell, S. E., Ahlbom, A., Erwald, R., Jacobson, G., Lindberg-Navier, I., Olin, R., Tornberg, B., and Wiechel, K. L. Diet and pancreatic cancer: A case-control study. *American Journal of Epidemiology,* 1986; 124(6): 894–902.

Olson, Rick. *Ionized Alkaline Water Using Platinum Electrolysis, Micro-Water and Coral Calcium* [proprietary marketing pamphlet for Coral Calcium]. Olympia, Wash.: Vitality Press and Product Information, September 1995.

Pasquale, A. D., Monforte, M. T., Calabro, M. L. HPLC analysis of oleuropein and some flavonoids in leaf and bud of *Olea europaea. Il Farmaco,* 1991; 46(6): 803–15.

Pearson, R. B. *Pasteur: Plagiarist, Impostor! The Germ Theory Exploded* (1942). Reprinted Pomeroy, Wash.: National Health Research Association. (See Resources section for information on National Health Research Association.)

————. *The Dream and Lie of Louis Pasteur.* Collingwood, Australia: Sumeria Press, 1994.

Peck, S. M., and Rosenfeld, H. The effects of hydrogen ion concentration, fatty acids and vitamin C on the growth of fungi. *Journal of Investigative Dermatology,* 1938; 1: 237–65.

Perlman, H. H. Undecylenic acid given orally in psoriasis and neurodermatitis. *Journal of the American Medical Association,* 1949; 139: 444–47.

Perlman, H. H., and Milberg, I. L. Peroral administration of undecylenic acid in psoriasis. *Journal of the American Medical Association,* 1949; 140: 865–68.

Peska, J. J., and Bondy, G. S. Alteration of immune function following dietary mycotoxin exposure. *Canadian Journal of Physiology and Pharmacology,* 1990; 68(7): 1009–16.

Rapaport, S. I. Blood coagulation and its alterations in hemorrhagic and thrombotic disorders. *The Western Journal of Medicine,* 1993; 158: 153.

Ren, A., and Han, X. Dietary factors and esophageal cancer: A case-control study. *Chinese Journal of Epidemiology,* 1991; 12(4): 200–4.

Rodricks, J. B., Hessiltine, C. W., Mehlman, M. A., eds. *Mycotoxins in Human and Animal Health.* Park Forest South, Ill.: Pathotox Publishers, 1977.

Rosenberg, E. W., Belew, P. W., Skinner, R. B., and Crutcher, N. Response to Crohn's disease and psoriasis. *New England Journal of Medicine,* 1983; 308: 101.

Saleem, A., et al. Viscoelastic measurement of clot formation: A new test of platelet function. *Annals of Clinical and Laboratory Science,* 1983; 13: 115.

Sandler, M., and Bourne, G. H., eds. *Atherosclerosis and Its Origin.* New York and London: Academic Press, 1963.

Sava, G., Giraldi, T., Mestroni, G., and Zassinovich, G. Antitumor effects of rhodium, iridium, and ruthenium complexes in comparison with cis-dichlorodiamino platinum in mice bearing Lewis lung carcinoma. *Chemico-Biological Interactions,* 1983; 45: 1–6.

Selig, M. S. Mechanisms by which antibiotics increase the incidence and severity of candidiasis and alter the immunological defense. *Bacteriological Review,* 1966; 30: 442–59.

Shook, E. E. *Advanced Treatise in Herbology.* Banning, Calif.: Enos Publishing Co., 1992.

Silberberg, J. M., et al. Identification of tissue factor in two human pancreatic cancer cell lines. *Cancer Research,* 1989; 49: 5443.

Spillert, C. R., et al. Altered coagulability: An aid to selective breast biopsy. *Journal of the National Medical Association,* 1993; 85: 273.

Sprince, H., et al. Protective action of ascorbic acid and sulfur compounds (including N-acetyl cysteine) against toxicity: Implications in alcoholism and smoking. *Agents and Actions,* 1975; 5: 164–73.

Steinmetz, K. A., and Potter, J. D. Food-group consumption and colon cancer in the Adelaide case-control study: Meat, poultry, seafood, dairy foods and eggs. *International Journal of Cancer,* 1993; 53(5): 720–27.

Structure, Betina V. Activity relationships among mycotoxins. *Chemico-Biological Interactions,* 1989; 71(2–3): 105–46.

Sugiyama, S., et al. The role of leukotoxin (9, 10-epoxy-12-octadecenoate) in the genesis of coagulation abnormalities. *Life Sciences,* 1988; 43: 221.

Tallman, M. S., et al. New insights into the pathogenesis of coagulation dysfunction in acute promyelocytic leukemia. *Leukemia and Lymphoma,* 1993; 11: 27.

Topley and Wilson. *Principles of Bacteriology, Virology and Immunity.* Baltimore: Williams & Wilkins, 1984.

Toth, B., and Gannett, P. Carcinogenesis study in mice by 3-methylbutanol methylformylhydrazone of *Gyromitra esculenta,* in vivo. *Mycopathologia,* 1990; 4(5): 283–88.

Toth, B., Patil, K., Erickson, J., and Kupper, R. False morel mushroom *Gyromitra esculenta* toxin: N-methyl-N-formylhydrazone carcinogenesis in mice. *Mycopathologia,* 1979; 68(2): 121–28.

Toth, B., Patil, K., Pyssalo, H., Stessman, C., and Gannett, P. Cancer induction in mice by feeding the raw morel mushroom *Gyromitra esculenta*. *Cancer Research,* 1992; 52(8): 2279–84.

Toth, B., Taylor, J., and Gannett, P. Tumor induction with hexanol methylformylhydrazone of *Gyromitra esculenta*. *Mycopathologia,* 1991; 115(2): 65–71.

Tranter, H. S., Tassou, S., and Nychas, G. J. The effect of the olive phenolic compound, oleuropein, on growth and enterotoxin B production by *Staphylococcus aureus*. *Journal of Applied Microbiology,* 1993; 74: 253–59.

Trousseau, A. Phlegmasia alba dolens. *Clinique Médicale de l'Hôtel-Dieu de Paris,* 1865, 3: 94.

Truss, C. Orian, M.D. *The Missing Diagnosis.* Birmingham, Ala.: The Missing Diagnosis, Inc., 1983.

Uraguchi, K., and Yamazaki, M., eds. *Toxicology, Biochemistry and Pathology of Mycotoxins.* New York: John Wiley & Son, 1978.

Van Deventer, S. J. H., et al. Intestinal endotoxemia. *Gastroenterology,* 1988; 94(3): 825–31.

Virchow, R. Hypercoagulability: A review of its development, clinical application, and recent progress. *Gesammelte Abhandlungen zur Wissenschäftlichen Medizin,* 1856; 26: 477.

Visioli, F., and Galli, C. Oleuropein protects low density lipoprotein from oxidation. *Life Sciences,* 1994; 55: 1965–71.

Wallach, Joel, B.S., D.V.M., N.D. *Rare Earths.* Bonita, Calif.: Ma Lan and Double Happiness Publishing Co., 1994.

White, A., et al., eds. *Principles of Biochemistry.* New York: McGraw-Hill Book Co., 1964.

Wilson, C. L. The alternatively spliced V region contributes to the differential incorporation of plasma and cellular fibronectins into fibrin clots. *Journal of Cell Biology,* 1992; 119: 923.

Wray, B. B., and O'Steen, J. M. Mycotoxin-producing fungi from house associated with leukemia. *Archived Environmental Health,* 1975; 30: 571–73.

Wray, B. B., Rushing, E. J., Schindel, A., and Boyd, R. C. Suppression of response to phytohemagglutinin in guinea pigs by fungi from a leukemia-associated house. *Archived Environmental Health,* 1979; 22: 400.

Wylie, T. D., and Morehouse, L. G. *Mycotoxic Fungi, Mycotoxins, Mycotoxicoses: An Encyclopedia Handbook,* Vol. 3.

Yamada, O., et al. Deleterious effects of endotoxins on cultured endothelial cells: An in vitro model of vascular injury. *Inflammation,* 1981; 5: 115.

Yoshida, S., Kasuga, S. H., Hayashi, N., Ushiroguchi, T., Matsura, H., and Nakagawa, S. Anti-fungal activity of garlic. *Applied and Environmental Microbiology,* 1987; 53(3): 615–17.

Young, Robert O. Fermentology and oxidology. The study of fungus-produced mycotoxic species and the activation of the immune system and release of reactive oxygen toxic species (ROTS) [Self-published]. Alpine, Utah: InnerLight Biological Research Foundation, 1994.

Zieve L., et al. Effect of hepatic failure toxins on liver thymidine kinase activity and ornithine decarboxylase activity after massive necrosis with acetaminophen in the rat. *Journal of Laboratory and Clinical Medicine,* 1985; 106(5): 583–88.

Zwicker, G. M., Carlton, W. W., and Tuite, J. Long-term administration of sterigmatocystin and *Penicillium viridicatum* to mice. *Food, Cosmetics and Toxicology,* 1974; 12: 491.